A series of student texts in

CONTEMPORARY BIOLOGY

General Editors:
Professor E. J. W. Barrington, F.R.S.
Professor Arthur J. Willis

The Comparative Endocrinology

of the Invertebrates

Kenneth C. Highnam
M.Sc., Ph.D.

Reader in Invertebrate Endocrinology,
University of Sheffield

and

Leonard Hill
B.Sc., Ph.D.

Lecturer in Zoology,
University of Sheffield

Ed

AMERICAN ELSEVIER PUBLISHING
COMPANY, INC.

52 Vanderbilt Ave., New York, N.Y. 10017

Printed in Great Britain by
William Clowes and Sons, Limited, London and Beccles

Preface

From its beginning endocrinology has been an experimental science. Yet because of legal restrictions (particularly in the United Kingdom) upon the use of vertebrate animals for experimental purposes, the undergraduate student has been forced to consider vertebrate endocrinology a largely theoretical subject, a confusing mass of text-book information to be absorbed and unravelled without the benefits of personal laboratory investigations.

In this book we have attempted to show that the principles of hormonal integration and co-ordination are as well exemplified in the invertebrate animals as in the vertebrate. These principles can be demonstrated practically, using species which are not precluded by the vivisection regulations, and which are often more suitable than vertebrates for class use. We have therefore incorporated a certain amount of practical detail into the text to illustrate the relatively simple experiments and techniques involved. It is hoped that this will encourage teachers and students to initiate their own programmes of endocrinological investigations.

Insects, in particular, are excellent animals to use for such experiments. Many species can be reared easily in the laboratory, and their endocrine systems, although simpler than those of any vertebrate, are sufficiently complex to pose suitably satisfying problems of technique and interpretation. It is not difficult to make extracts of the corpora cardiaca of a locust or a cockroach and to test their effects upon the rate of heart beat or upon diuresis. But then further experiments are necessary to decide whether the activities of the extracts are due to the intrinsic hormones of the corpora cardiaca, or to hormones produced elsewhere and stored temporarily in the glands. Ecdysone, the insect moulting hormone, is now available commercially, together with plant steroids such as ponasterone and inokosterone which mimic its effects. Other compounds with juvenile hormone activity can also be purchased. The effects of these substances upon the development and reproduction of laboratory colonies of insects can thus be tested, even when facilities are not available for surgical operations upon the endocrine systems of the insects. The large proportion of this book devoted to insect endocrinology reflects not only the amount of information on the subject now available, but also our feeling that laboratory investigations on insect hormones should be of considerable importance in view of the current emphasis upon experimental method in biology in both schools and universities.

Neurosecretory hormones dominate this account of invertebrate endocrinology. This is inevitable when many of the groups appear to possess only this particular class of hormones. But more importantly, it is our opinion that in almost all animals neurosecretory mechanisms are of the greatest significance in relating the development and reproduction of an individual to its changing environment, and in this book we have tried to give reasons for this view. The suggestion that vertebrate animals are no different from the invertebrates in this regard explains the presence of some vertebrate endocrinology within our pages. This concept of the role of neurosecretion owes a very great deal to the writings of Professors Ernst and Berta Scharrer.

We have assumed throughout some knowledge of the structural organization and biology of the animals whose endocrinology is discussed, although more detail has been given where it was considered necessary. Other volumes in this series should do much to make up any deficiencies.

We gratefully acknowledge the generosity of Dr. W. Mordue and Dr. G. Goldsworthy in allowing us to quote data on diuretic and hyperglycaemic hormones in the locust (Chapter 8) prior to their publication. We are also grateful to Mr. M. West for the preparations from which Plates 5 and 6 were made, and to Mr. R. A. Johnson for providing Fig. 12.11. Our thanks are further due to Mr. R. Webb who made the photomicrographs.

Sheffield,
1969

K.C.H.
L.H.

Table of Contents

I

Nervous and Chemical Co-ordination

In all Metazoa, the nervous and endocrine systems so co-ordinate the activities of the various organs and tissues in the body that the animals function as individuals. The nervous system serves for rapid communication, essential, for example, in the systematic contractions of muscles during locomotion. For this, complicated chains of interconnected neurones are necessary for the transmission of transient impulses, together with the highly localized production of chemicals such as adrenaline and acetylcholine which are rapidly destroyed. The endocrine system uses circulating body fluids to carry its chemical messengers to more or less specific target organs. These chemicals take time to build up to an effective concentration, and consequently must have a longer biological life than the chemicals of the nervous system before they are eventually destroyed or excreted. Hormones are consequently well suited to exert their effects over extended periods of time, and the endocrine system controls long-term processes within the body, such as the co-ordinated growth of organs or the maintenance of appropriate metabolite concentrations in the blood and tissues.

An animal derives information about the chemical and physical features of its environment through sensory receptors. This information passes to the nervous system where it is collated and interpreted, and motor impulses are initiated which cause the animal to react appropriately. Short-term environmental changes, such as the appearance of predators or prey, the scents produced by possible mates, or the hazards associated with a sudden downpour of rain, each evoke rapid and particular responses in individuals through nervous activities. But such reactions are not necessarily rigidly fixed. If the stimuli are repeated many times, the sensory receptors may adapt and no longer respond. Or the animal may habituate to the stimuli, learning to disregard them. The same stimulation may evoke different reactions when the animal is feeding, moving or engaging in courtship. Moreover, a particular activity can have a marked after effect when it is completed, considerably modifying the immediately subsequent behaviour.[168, 169] Behavioural variability is the result of complicated neuronal interactions within the central nervous system.

It must not be supposed that the nervous system functions quite independently of the endocrine system. A mature female grasshopper will move towards and mate with a courting male. An immature female avoids him, and will even fight him if he persists in his courtship.[193] The hormone complements of the two females cause their central nervous systems to interpret the same information differently, and to initiate quite opposite reactions. It is likely that central nervous activity in most animals is strongly affected by hormones most of the time.

The opposite is also true: hormone production and release is dependent upon nervous activity. Nearly all animals have to respond developmentally to environmental changes, particularly seasonal fluctuations, throughout the year. Unfavourable periods are avoided by dormancy or migration, or overcome by other changes in habit or physiology. Advantage is taken of favourable seasons for rapid development and multiplication. Even transient fluctuations, such as temporary shortage of food, or the absence of suitable mates, can have dramatic effects upon development. The act of mating in a female insect can accelerate the development of her eggs; the changing length of day can control the onset of metamorphosis in annelids; colour change in many crustaceans is brought about by alterations in the shade of their immediate environment. All these developmental or physiological responses result from changes in the concentrations of circulating hormones caused by nervous impulses originating in the stimulation of particular sense organs. The nervous and endocrine systems are strictly interdependent: the normal activity of either requires the functional presence of the other.

Paradoxically, there are very few endocrine organs which are richly innervated. The adrenal medulla of the Vertebrates is an obvious exception, innervated by pre-ganglionic fibres from the coeliac plexus, part of the sympathetic nervous system. But the cells of the adrenal medulla are considered to be modified post-ganglionic sympathetic neurones, specialized for the production of large amounts of adrenaline and noradrenaline—the chemicals produced by normal post-ganglionic neurones —to be poured into the circulation at the appropriate times of fight or flight. The adrenal medulla is consequently a rather atypical endocrine gland. Other endocrine organs in vertebrate animals may contain nerve fibres, but these are usually associated with the blood supply to the glands. Endocrine activity may be partly controlled by altering the blood flow through glands, but the mechanism is not sufficiently all-embracing to explain the often close correlation between hormonal processes and environmental events. In particular, the anterior pituitary has been long known to produce a number of hormones which control the activities of other endocrine glands, but it contains very few nerve fibres indeed.

How then can nervous activity control endocrine function? More than fifty years ago, Stefan Kopec suggested that the larval brain of the gipsy moth, *Lymantria dispar*, produced a hormone which induced pupation.[177] At this time, both nervous and endocrine mechanisms, particularly in vertebrate animals, were being intensively investigated. For many reasons, these researches followed quite separate paths and it was never seriously considered that nervous systems could produce hormones. Kopec's experiments attracted very little attention, especially when it was subsequently shown that in other insects epithelial endocrine glands in the head, the corpora allata, apparently played an important role in the control of moulting and metamorphosis (Chapter 6). In the butterflies and moths, the corpora allata lie close to the brain, and it was assumed that Kopec had removed these glands unintentionally, together with the brain.

But during the last twenty years, opinions about the possible endocrine function of nervous tissue have been completely reversed. It is now generally accepted that most, if not all, nervous systems *do* have the ability to produce hormones. This new point of view results from the experimental demonstration that the central nervous systems of arthropods and vertebrates contain particular endocrine cells. These cells are morphologically similar to neurones, with axons, dendrites, Nissl granules and neurofibrillae. They are also able to transmit nervous impulses, but they differ from other neurones in two important respects: their axons do not innervate effector organs such as muscles, nor make synaptic connections with other neurones; and they manufacture materials, often visible in stained sections of nervous tissue, which are released from the ends of the axons and exert a biological effect some distance away. In effect, the cells are neurones which also produce hormones and are consequently called neurosecretory cells (Fig. 1.1).

In many animals, the neurosecretory cells are often clumped into groups which are very conspicuous features of the central nervous system after appropriate staining. The ends of the neurosecretory cell axons are usually swollen, and the material manufactured by the cell can be stored here before being released. The swollen axon terminals lie outside the nervous system, usually closely associated with the circulatory system

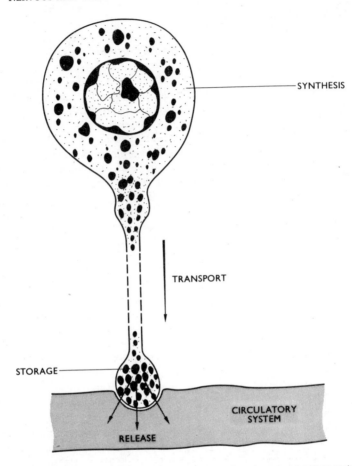

Fig. 1.1 Diagram of a neurosecretory cell. Globules of neurosecretory material, which can be stained and made visible with the light microscope, are synthesized in the cell body, or perikaryon, and transported along the axon to the swollen terminal where they are stored or released into the circulatory system. In vertebrate animals, it is likely that synthesis can also occur within the axon.

which carries the neurosecretory hormones around the body. Where numbers of neurosecretory cells are grouped together, their axon terminals can form well developed organs outside the nervous system, and since these structures are combined with blood vessels, they are called *neurohaemal organs*.

Neurosecretory cells from invertebrates and vertebrates are essentially similar, although it is possible that the invertebrate cells may lack dendrites and it is not yet *conclusively* proved that they can transmit nervous impulses.[266, 293] The material elaborated by neurosecretory cells is largely protein and reacts histologically and histochemically in much the same way whether it is produced by invertebrate or vertebrate neurosecretory cells. Ultrastructurally, the resemblance is even closer: the neurosecretory particles are usually electron-dense spheres between 1000 to 3000 Å in diameter, surrounded by a thin membrane (Fig. 1.2). Some

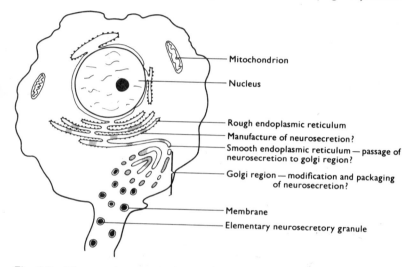

Fig. 1.2 Diagram of part of the ultrastructural organization of a neurosecretory cell. Protein, which makes up at least part of each elementary neurosecretory particle, is synthesized upon ribosomes associated with the endoplasmic reticulum. The protein material is passed to the Golgi apparatus where it is organized and provided with a membrane covering.

neurosecretory cells in both invertebrates and vertebrates produce electron-transparent vesicles. The dense spheres originate in the Golgi apparatus of the cell bodies,[14] in a manner similar to protein droplet formation in other endocrine and ordinary gland cells. But in addition, synthesis, or perhaps reorganization, of neurosecretory material can take place in the axons of many neurosecretory cells. The protein material synthesized by neurosecretory cells, called **neurophysine** in the vertebrates, is considered to be a carrier for the actual hormones. There is good evidence for this from vertebrate neurosecretory systems; whether the same applies to all invertebrate neurosecretions is problematical.

The discovery of neurosecretory cells has added a new dimension to the study of endocrine systems. Previously, endocrine glands were divided into two categories: those which were derived from embryonic ectoderm, mesoderm and endoderm layers, called epithelial endocrine glands; and those which were derived from nervous tissue. The first category included all the endocrine glands and tissues of a typical vertebrate, except the adrenal medulla and the posterior pituitary which developed from a transformed sympathetic ganglion and part of the brain respectively, and were therefore placed in the second category. But it is now known that the posterior pituitary is a neurohaemal organ for groups of neurosecretory cells in the hypothalamus of the brain. The posterior pituitary hormones—all octapeptides like the oxytocin and vasopressin of mammals—are manufactured by the hypothalamic neurosecretory cells and released into the blood from the pars nervosa, perhaps after preliminary storage there (Fig. 1.3). The posterior pituitary is thus quite a different endocrine organ from the adrenal medulla. Other neurosecretory hormones are released from the median eminence into the

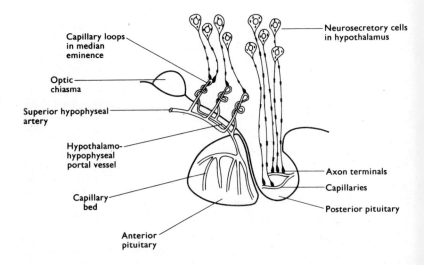

Fig. 1.3   The relationships between neurosecretory cells in the hypothalamus of the brain and the pituitary in vertebrates. Some neurosecretory cells send their axons directly to blood capillaries within the posterior pituitary, and release the octapeptides oxytocin and vasopressin. Other neurosecretory cell axons terminate upon capillaries in the anterior part of the infundibular stalk, the median eminence. These capillaries pass to the anterior pituitary, carrying blood containing neurosecretory factors which control the release and production of anterior pituitary hormones. Compare Plate 3.

capillaries of the hypothalamo-hypophyseal portal system[66] (Fig. 1.3). These **releasing factors** are carried to the anterior pituitary to control the release and synthesis of the anterior pituitary hormones.[123] In some vertebrates a second major neurosecretory system is situated at the posterior end of the spinal cord.[103] The neurohaemal organ for this caudal neurosecretory system is the urophysis:[15] the functions of the hormones produced here have not so far been precisely determined.[16, 292]

In the invertebrate animals, epithelial endocrine glands, endocrine organs developed from transformed nerve ganglia, and neurosecretory cells often with well formed neurohaemal organs are all present, though not necessarily together in the same taxonomic group (Chapter 2). Even in complicated animals like the arthropods, the number of epithelial endocrine glands is much less than in the vertebrates, and in most other invertebrate groups they are absent altogether. Endocrine mechanisms in the invertebrates are consequently usually simpler than the intricate interactions between numerous endocrine glands producing even larger numbers of hormones which characterize vertebrate endocrinology. Because of the relative paucity of epithelial endocrine glands in the invertebrates, neurosecretory mechanisms assume great importance.

Their dual properties make neurosecretory cells the most likely candidates for transforming nervous into endocrine information.[236] Where this function is not complicated by the presence of numerous other endocrine mechanisms, the significance of neurosecretory processes in co-ordinating developmental events within an animal with environmental fluctuations may be more clearly demonstrated. The vertebrate hypothalamo-hypophyseal neurosecretory system plays the same part as the simpler neurosecretory systems in the invertebrate animals: a knowledge of these systems must surely precede any study of the much more exact but more complicated mechanisms found in the vertebrates.

However, many problems are associated with the determination of the functions of neurosecretory systems. The classical endocrinological experiment involves the removal of a suspected endocrine gland followed by its subsequent reimplantation to a different site in the body. If the consequences of removal are reversed and brought back to normal on reimplantation, then it is established that a hormonal mechanism is involved. But when a neurohaemal organ is removed, the cut ends of the neurosecretory axons remaining within the animal may still release hormones, sometimes in an uncontrolled manner. Or a new neurohaemal organ may be rapidly regenerated, so that the effects of a deficit of neurosecretory hormones may not be obvious. Moreover, the reimplantation of the neurohaemal organ may introduce several hormones into the animal in concentrations, and in relative proportions, which are never normally attained. When the neurosecretory cells which supply the neurohaemal organ are destroyed or removed, the stored hormones with-in the organ may be released, or may leach out, for some time after the operation. Again, hormonal deficiency may not be immediately apparent. In some animals, the neurosecretory cells may be scattered singly or in small numbers throughout the central nervous system (Chapter 2), and a well defined neurohaemal organ may be lacking. Some neurosecretory hormones are not released to the circulatory system, but are transported axonally directly to their target organs.[156] It is difficult to determine the exact function of such neurosecretory mechanisms because of the im-possibility of extracting and testing biological materials from individual cells. In many instances, the neurosecretory nature of such cells is in-ferred simply because the cells are histologically and histochemically similar to those in other animals whose function has been determined experimentally.

The invertebrates include groups which possess epithelial endocrine glands whose functions can be elucidated by the classical techniques of extirpation and reimplantation (Chapter 2). A very small number of hormones from such glands have been purified, their chemical composi-tion determined and they can now be synthesized (Chapter 12). This allows much more sophisticated experiments in which the effects of different concentrations can be established, the ways in which combina-tions of hormones can interact, how the introduction of particular hor-mones at unnatural times can affect development, and so on. The well developed neurosecretory systems of some invertebrates allow detailed investigations into the problems of the production and release of com-binations of neurosecretory hormones. Other invertebrates possess neurosecretory mechanisms whose functions can only be inferred from circumstantial evidence. In subsequent chapters, an attempt will be made to analyse the evidence for endocrine function in a broad range of in-vertebrate animals. Although the evidence may be slight in many instances, it would be unreasonable to exclude it: it could provide the foundation for much fruitful research.

2

Invertebrate Endocrine Systems

THE COELENTERATA, TURBELLARIA, NEMERTEA AND NEMATODA

In the less highly organized invertebrates, up to and including the annelids and molluscs (but with the exception of the cephalopods p. 15) epithelial endocrine glands are apparently absent. Neurosecretions therefore are the only hormonal co-ordinators. In *Hydra* (Coelenterata) neurosecretory cells are present in the nerve net of the hypostome (Fig. 2.1) and release their contents into the mesogloea.[29] *Polycelis nigra* (Turbellaria) has many neurosecretory cells in the ventral part of the brain, increasing in number in the more posterior regions.[189] In the Nemertea, a curious structure called the ***cerebral organ***, composed of glandular and neural elements, has for long been compared with neurosecretory structures in other groups.[232] In the more primitive Palaeonemertea, the cerebral organ is epithelial in position, connected with the cerebral ganglion by a long nerve (Fig. 2.2a). In more advanced members of the group, the cerebral organ has migrated internally, retaining its connection with the epithelium by the cerebral canal (Fig. 2.2b) while in the Heteronemertea, the cerebral organ is fused with the cerebral ganglion (Fig. 2.2c). This progressive incorporation of a glandular structure with a central nervous system was thought at one time to suggest the way in which neurosecretory structures originated (see p. 29). But recently, more typical neurosecretory cells have been found in the cerebral ganglion of several heteronemertean species, which have no connection whatsoever with the cerebral organ.[186] Within the Nematoda,

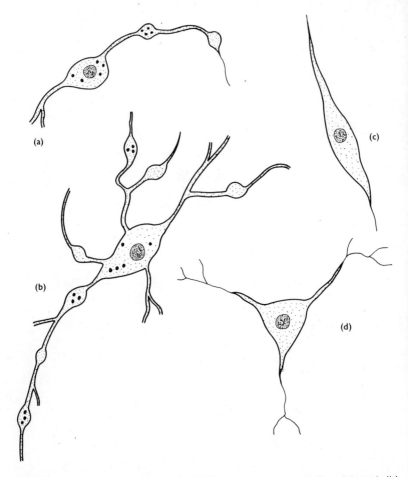

Fig. 2.1   Neurones from the sub-hypostomal region of *Hydra*. (a) and (b) contain stainable granules in the perikaryon and within vesicles arranged along the dendritic processes. Some vesicles are devoid of granules. (c) and (d) are bi- and tri-polar neurones respectively, containing no vesicles or granules. (After Burnett and Diehl[29])

Ascaris lumbricoides has neurosecretory cells in the lateral ganglia on the anterior nerve ring (Fig. 2.3), and *Phocanema decipiens* has similar cells in the dorsal and ventral ganglia[74] (Fig. 2.4). Many of the primary sense organs on the lips of these two species also stain in the same way as the ganglionic neurosecretory cells.

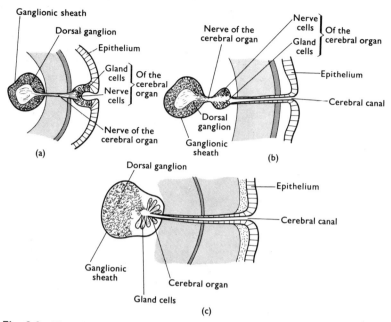

Fig. 2.2 The relationships between brain and cerebral organ in three nemerteans. In (a) the glandular cerebral organ lies in the surface epithelium of the body, connected with the dorsal ganglion by a long nerve. In (b) the cerebral organ is internal, connected to the dorsal ganglion by only a short nerve, with a long canal leading externally. In (c) the gland cells of the cerebral organ lie within the ganglionic sheath, although still connected with the exterior by the long cerebral canal. (After Scharrer[232])

Experiments to determine the functions of these neurosecretory cells have been carried out in all these less well organized invertebrates (Chapter 3). But because of the relatively small numbers of cells present, their dispersion within the nerve ganglia, and the complete absence of any neurohaemal organs, or their analogues, in practice it is impossible to relate any experimental results precisely to factors produced by the neurosecretory cells.

THE ANNELIDA

Groups of scattered neurosecretory cells are found in the brain, or supra-oesophageal ganglion, and in the ganglia of the ventral chain in both polychaetes and oligochaetes[100] (Fig. 2.5). In some polychaetes,

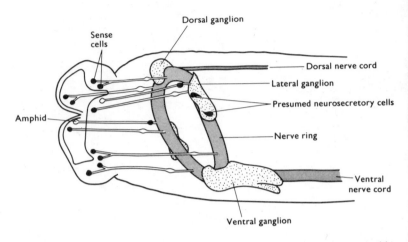

Fig. 2.3 The anterior nervous system of the nematode *Ascaris lumbricoides.* Neurosecretory cells (shown in black) are present in the lateral ganglia on each side of the nerve ring. Many of the sense cells on the lips also stain in the same way as neurosecretory cells. (After Davey [74])

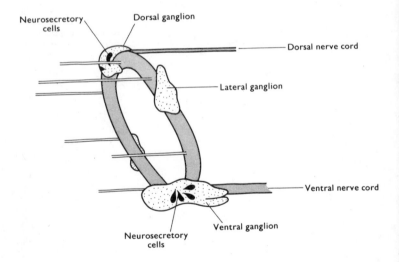

Fig. 2.4 The anterior nervous system of the nematode *Phocanema decipiens.* In this species neurosecretory cells (shown in black) occur in the dorsal and ventral ganglia of the nerve ring. (After Davey [74])

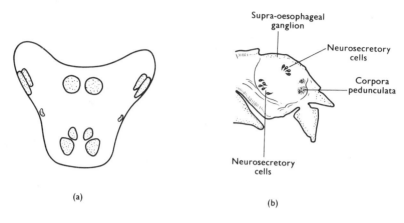

Fig. 2.5 (a) Plan of the brain of *Nereis diversicolor* showing the ganglionic nuclei (stippled) containing neurosecretory cells which are reported to show histological changes after the amputation of posterior segments of the body. (b) Parasagittal section through the prostomium and brain of *Platynereis dumerilii*. Two major groups of neurosecretory cells are present in the brain which are likely to be the source of the oocyte inhibiting hormone. ((a) After Clark and Bonney[57]; (b) after Hauenschild[125])

neurosecretory axon tracts extend ventrally through the brain to terminate in a specially modified part of the pericapsular membrane in the region of the dorsal blood vessel[54, 21] (Fig. 2.6). This structure is possibly a simple neurohaemal organ, but if so, only a small proportion of the neurosecretory axons terminate there.

It is likely that many so-called neurosecretory cells in the annelid nervous system are not neurosecretory at all. In some leeches, 50 to 95% of the neurones in the brain have been called neurosecretory cells.[119] It must be realized that the criteria for identifying a neurosecretory cell are often solely histological: the presence of inclusions within the perikarya, the amount varying under different conditions; inclusions along the axons from the cells; and if present, a neurohaemal organ in which the axons terminate. But many ordinary nerve cells may contain inclusions which vary in amount at different times, and may duplicate the staining properties of neurosecretory proteins. Even ultrastructurally, some neurones contain membrane-bound elementary particles, 1000–3000 Å in diameter, which resemble closely elementary neurosecretory granules; recently electron-transparent neurosecretory vesicles have been described which are difficult to differentiate from synaptic vesicles found at axonal synaptic junctions. The only true test of neurosecretory function is the experimental demonstration of endocrine activity: the annelid brain passes this test (p. 47). But even so, endocrine function cannot be attributed to particular neurosecretory cell groups in the brain.

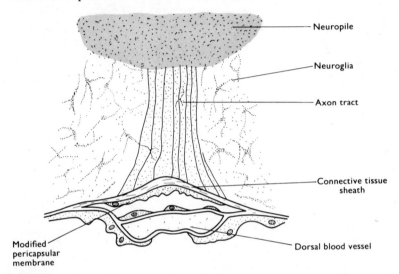

Neuropile

Neuroglia

Axon tract

Connective tissue sheath

Modified pericapsular membrane

Dorsal blood vessel

Fig. 2.6 Cross section of the ventral part of the brain of *Nephtys californiensis* showing the pericapsular organ, an area where neurosecretory cells in the posterior part of the brain may release their secretions into the blood. It is by no means certain that *all* the axons from the posterior groups of neurosecretory cells run to this cerebro-vascular complex. There are also anterior groups of neurosecretory cells in the brain which have no connection with the organ; their secretions must be released elsewhere. (After Clark[54])

THE MOLLUSCA

Considering the size and diversity of the phylum, studies on neurosecretion in the Mollusca have been remarkably limited. But at the least, it has become clear that no well defined endocrine system exists in the majority of the molluscs studied. With the exception of the cephalopods, there is no clear evidence that any group of molluscs possesses any epithelial endocrine glands.

Neurones with the histological characteristics of neurosecretory cells occur in almost all ganglia of the nervous system in gastropods and lamellibranchs (Figs. 2.7, 2.8). Histological changes in these cells have been correlated with a variety of developmental events (p. 67). But

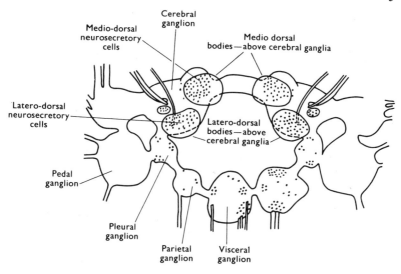

Fig. 2.7 The central ganglia of the pond snail, *Lymnaea stagnalis,* showing the location of scattered neurosecretory cells (black dots). The positions of the medio-dorsal and latero-dorsal bodies above the cerebral ganglia are also shown. (After Lever, de Vries and Jager[190])

whether such cells are actually neurosecretory is still open to question: many of the stainable inclusions could be lysosomes, or glycogen granules, pigmented droplets, etc. (Chapter 5).

Particular regions of the perineurium of nerves adjacent to blood vessels, containing much neurosecretory material (Fig. 2.9), occur in some gastropod molluscs. These could represent the simplest of all neurohaemal organs. But definite experimental evidence that neurosecretory hormones are stored in these areas is still lacking. Neurosecretory cells are also said to occur in the tentacles of some pulmonate gastropods, and there is evidence that these cells exert some developmental effect (p. 70).[220, 180] But unfortunately, the neurosecretory nature even of these cells is not definitely proved.

In the cephalopod molluscs (octopus and squids), the optic glands are well developed epithelial endocrine organs lying on the optic stalks just internal to the eyes (Fig. 2.10). These glands are intimately involved in the control of reproductive development (p. 73). Although neurosecretory mechanisms *may* be involved with the function of the optic glands (Fig. 2.10), no direct experimental proof is yet available. The mesodermal branchial glands on the gills (Fig. 2.11) are also endocrine

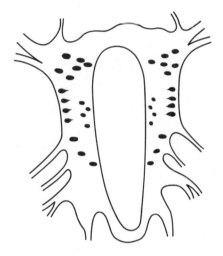

Cerebropleural ganglion

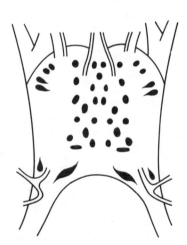

Visceroparietal ganglion

Fig. 2.8 Positions of cells in the cerebropleural and visceroparietal ganglia of a lamellibranch mollusc which give a positive reaction with stains which are used to characterize neurosecretory cells in other animals. These scattered cells may therefore be neurosecretory, although there is little experimental evidence to confirm this view. (After Antheunisse[3])

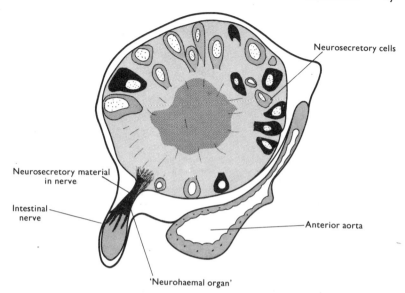

Fig. 2.9 Cross section through the visceral ganglion and intestinal nerve of a gastropod mollusc. The axons of neurosecretory cells in the ganglion pass into the intestinal nerve, where their close association with the anterior aorta has suggested that this area of the nerve may be a neurohaemal organ. (After Simpson, Bern and Nishioka[246])

organs, said to have functions similar to those of the adrenal glands of vertebrates (p. 75).

THE ARTHROPODA

The arthropods are complicated, highly organized invertebrate animals. In number of species, and in number of individuals, they are the most successful of all animal groups in their colonization of land and sea. The insects and the crustaceans are pre-eminently the dominant groups among the arthropods. With the exception of the cephalopod molluscs (also highly organized animals), the arthropods are the only invertebrates known to possess epithelial endocrine glands. In combination with well-developed neurosecretory mechanisms, the epithelial endocrine glands form endocrine systems which in arrangement and scope resemble those of the vertebrate animals.

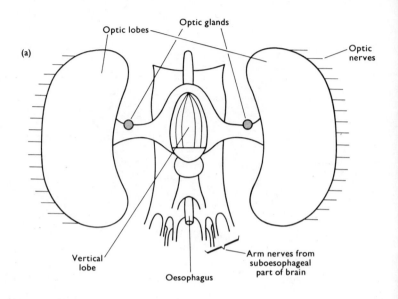

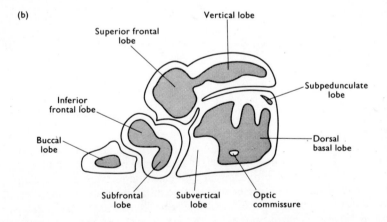

Fig. 2.10   Structure of the brain of *Octopus*. (a) Dorsal view of the brain as it would be seen after its removal from the cartilaginous brain case. The optic glands are situated dorsally on the optic stalks. (b) A longitudinal vertical section of the supra-oesophageal part of the brain showing its different areas. *For further details see text*. (After Wells and Wells[268])

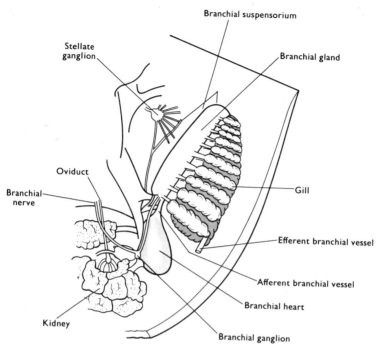

Fig. 2.11 The left branchial gland of the octopus as seen in a dissection from the ventral side. Notice the very close association of the gland with the gill and the branchial blood supply. (After Taki[258])

Insecta

The insect endocrine system has four major components: groups of **neurosecretory cells** in the brain, the **corpora cardiaca**, the **corpora allata**, and the **thoracic glands** or their equivalents (Fig. 2.12). The corpora allata and thoracic glands are epithelial endocrine glands, arising embryonically as ectodermal invaginations in the region of the first and second maxillary pouches.[154, 265, 213, 130] The corpora allata are connected by nerves to the corpora cardiaca and to the sub-oesophageal ganglion; the thoracic glands are innervated variously from the sub-oesophageal ganglion and thoracic ganglia (Figs. 2.12, 2.13). Structures homologous with the thoracic glands, the **ventral glands**, are found in some insects: they remain in the head after invagination instead of migrating further backwards into the thorax.[255] But in the cockroaches, both ventral and thoracic glands are present, so the structures must be serially homologous.[226, 227] The corpora cardiaca develop embryonically as invaginations of the foregut at the same time and in the same series as

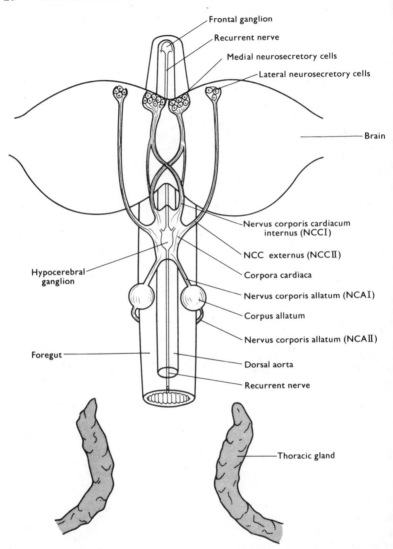

Fig. 2.12 Diagram of the insect endocrine system. Two groups of neuro-secretory cells in each half of the brain send axons to the corpora cardiaca, in close association with the aorta. The majority of the axons from the medial groups of neurosecretory cells decussate before leaving the brain. The corpora cardiaca are connected by nerves with the corpora allata, which also send nerves to the suboesophageal ganglion. The thoracic glands, or their equivalents, are not normally in anatomical connection with the corpora cardiaca and corpora allata.

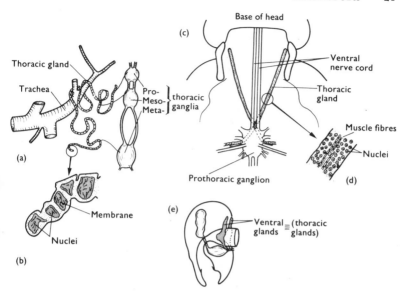

Fig. 2.13 Thoracic glands of the lime hawk moth (a, b), cockroach (c, d) and locust (e). The glands may be innervated from the thoracic ganglia (a, c) although in some insects the glands are found in the head and not the thorax, and are called ventral glands (e). The histology of the glands is variable : in the lime hawk moth, the nuclei are large and deeply folded with little surrounding cytoplasm (b); in the cockroach, the nuclei are small and spherical separated by considerable cytoplasm, and the glands contain small muscle fibres (d). ((c) After Rae and O'Farrell[227])

the dorsal sympathetic nervous system, which includes the frontal, hypocerebral and stomachic ganglia, connected by the recurrent nerve[229] (Fig. 2.14). The corpora cardiaca are consequently transformed nerve ganglia, and are still often referred to as oesophageal ganglia. As their name implies, they are usually in close association with the dorsal heart. The corpora cardiaca produce their own intrinsic hormones (Chapter 8), but their major function in most insects is to store and release neurosecretory hormones from the brain (Fig. 2.14). They are thus the most important insect neurohaemal organ; with the neurosecretory cells in the brain, they form the *cerebral neurosecretory system.* The cerebral neurosecretory cells are in two groups in each half of the pars intercerebralis region of the forebrain.[136] The medial groups are often closely apposed: most of their axons decussate in the brain before passing to the corpora cardiaca (Fig. 2.12). The lateral cell groups are usually smaller than the medial; their axons run directly to the ipsilateral corpus cardiacum. The corpora cardiaca thus receive paired inner and outer

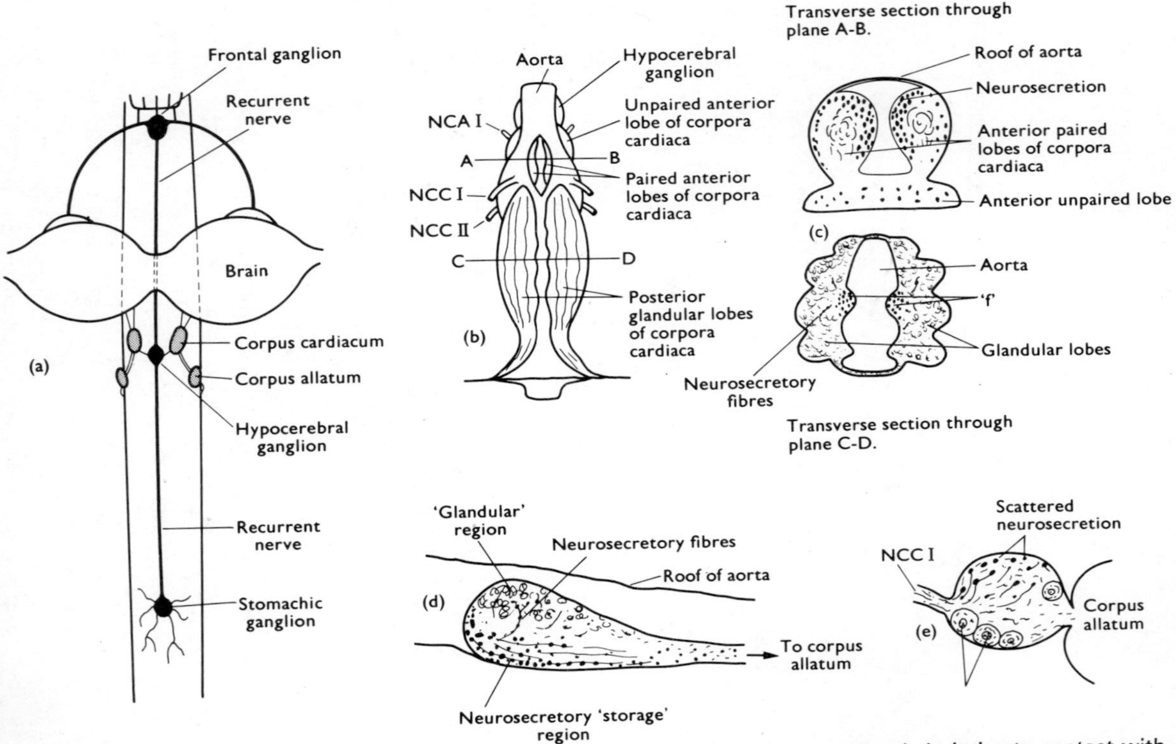

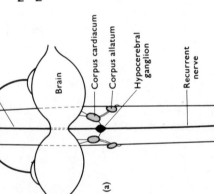

Fig. 2.14 (a) Relationships between the dorsal sympathetic nervous system (in black), corpora cardiaca and allata (cross hatched) and brain in a generalized insect. (b) corpora cardiaca of the locust, with the aorta cut open to show the anterior lobes. Note the separation of neurosecretory storage from glandular lobes of the corpora cardiaca. (c) Transverse sections through neurosecretory storage lobes and glandular lobes of the locust corpora cardiaca. The glands compose the greater part of the aorta wall in this region. The neurosecretory storage lobes send axons (marked 'f') posteriorly to the glandular lobes, and neurosecretory fibres ramify between the cells of the latter. (d) Longitudinal section through a corpus cardiacum of the cockroach. In contrast to the arrangement in the locust, the glandular region is in intimate contact with the neurosecretory storage region although neurosecretory axons still ramify between its cells. (e) Corpus cardiacum of the lime hawk moth, with large intrinsic cells and scattered neurosecretion.

Labels in figure:

(a) Frontal ganglion, Recurrent nerve, Brain, Corpus cardiacum, Corpus allatum, Hypocerebral ganglion, Recurrent nerve, Stomachic ganglion

(b) Aorta, Hypocerebral ganglion, Unpaired anterior lobe of corpora cardiaca, Paired anterior lobes of corpora cardiaca, Posterior glandular lobes of corpora cardiaca, Neurosecretory fibres, NCA I, NCC I, NCC II, A, B, C, D

(c) Transverse section through plane A-B: Roof of aorta, Neurosecretion, Anterior paired lobes of corpora cardiaca, Anterior unpaired lobe. Transverse section through plane C-D: Aorta, 'f', Glandular lobes

(d) Roof of aorta, Neurosecretory fibres, 'Glandular' region, Neurosecretory 'storage' region, To corpus allatum

(e) Scattered neurosecretion, Corpus allatum, NCC I

nerves from the cerebral neurosecretory cells. The size and staining re-actions of the medial cells allow the differentiation of four types, called A-, B-, C- and D-neurosecretory cells.[131] It is difficult to ascribe parti-cular functions to these cell types. Non-neurosecretory axons run with those from the neurosecretory cells to the corpora cardiaca: the neuronal origin of these fibres is unknown.

Other neurosecretory cells occur in the brain and in all the ventral ganglia of the central nervous system[76] (Fig. 2.15). Those in the ventral

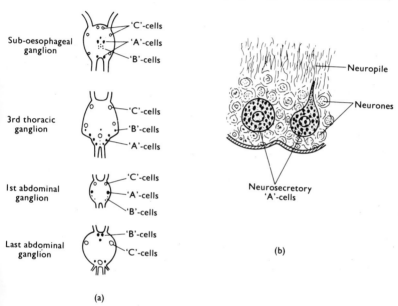

(a)

(b)

Fig. 2.15   (a) Arrangement of individual neurosecretory cells with different staining properties within some of the ventral ganglia of the locust (after Delphin[76]). (b) Drawing of two neurosecretory cells from the third thoracic ganglion of the locust.

ganglia are associated with a series of small neurohaemal organs, called the *perisympathetic organs*,[78] found in the vicinity of the small median ventral nerve (Fig. 2.16). With few exceptions (pp. 146, 158, 167) the functions of these cells are unknown.

In many groups of insects the originally paired corpora cardiaca and allata fuse to form single organs. A variety of arrangements of the glands is thus possible[43] (Fig. 2.17). Since the glands originate on either side of the aorta, this vessel is usually closely associated with the fused glands

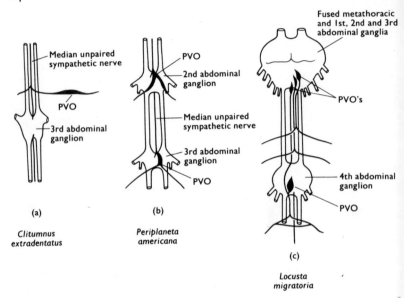

Fig. 2.16 Periviseral neurohaemal organs (PVO) associated with the ventral ganglia and ventral sympathetic nervous system of a stick insect, a cockroach, and a locust. ((a) After Raabe[78] (b) after de Besse[17a], (c) after Chalaye[44a])

(Fig. 2.17). The greatest degree of condensation of the endocrine system is found in larvae of the Diptera Cyclorrhapha (blowflies, houseflies, fruitflies, etc.) where there is a single corpus allatum dorsal to the aorta, a single corpus cardiacum ventral to the vessel, and the nerves connecting the gland around the sides of the aorta are surrounded by the homologues of the ventral glands (Fig. 2.18). The complex ring-shaped structure is called Weismann's ring after its discoverer.[267]

Neurosecretory cells in the central nervous system, corpora cardiaca and allata, and the thoracic glands all originate from ectodermal tissues in the embryo. Recently, the apical tissue of the testes in the glow-worm, *Lampyris noctiluca*, has been shown to produce a hormone[211] (p. 103). Such tissue is mesodermal in origin, as are the pericardial cells around the heart which may produce hormones in some insects (p. 164).

Crustacea

The crustacean endocrine system cannot be categorized so easily as that in the insects. There is a pair of **Y-organs** in the head, which are epithelial endocrine glands analogous to the insect thoracic glands, and in male crustaceans, a pair of **androgenic glands** in the vicinity of the sperm ducts[46] (Fig. 2.19). There are three major neurohaemal organs:

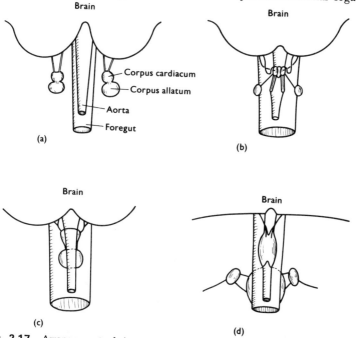

Brain

Corpus cardiacum
Corpus allatum
Aorta
Foregut

(a)

Brain

(b)

Brain

(c)

Brain

(d)

Fig. 2.17 Arrangement of the corpora cardiaca and corpora allata in (a) lime hawk moth, (b) locust, (c) *Pyrrhocoris apterus* (Hemiptera) and (d) *Ephemera danica*. *Ephemera* is unusual in that the major allatal nerves are those to the sub-oesophageal ganglion. Compare Fig. 2.12.

the **sinus glands**, in the eyestalks of the stalked-eyed crustacea and within the head of sessile-eyed forms; the **post-commissural** organs, behind the tritocerebral commissure; and the **pericardial organs**, in the vicinity of the heart (Fig. 2.19). The greatest confusion has arisen over the nomenclature of the neurosecretory cells which supply the sinus glands.[118, 41] Many of the cells are within the optic ganglia and are called the **X-organs**, but at least four X-organs, located in different segments of the optic ganglia have been described (Fig. 2.20). Other neurosecretory cell groups in the brain send axons to the sinus gland (Fig. 2.19). The post-commissural organs are innervated by neurosecretory cells in the tritocerebrum of the brain; the pericardial organs by at least seven cell groups in the thoracic ganglionic mass.

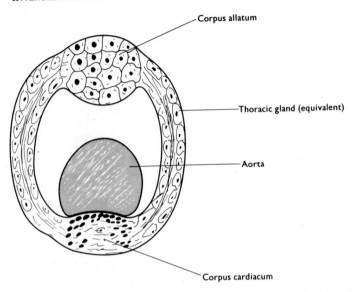

Fig. 2.18 Section through Weismann's ring of a blowfly larva, showing the single corpus allatum dorsal to the aorta, corpus cardiacum ventral to the vessel and the thoracic gland homologues connecting the two.

Other Arthropoda

Protocerebral neurosecretory cells are present in both chilopods and diplopods in the Myriapoda. In these groups, neurohaemal organs analogous to the insect corpora cardiaca and the crustacean sinus glands are also present: these are the ***cerebral glands***[225] (Fig. 2.21). A moulting centre behind the head in *Lithobius forficatus*, controlled by the cerebral glands, suggests the presence of analogues of the insect thoracic glands.[239] Many arachnids also possess cerebral neurosecretory cells, with a variety of neurohaemal organs where neurosecretions are stored and released.[98] In some arachnids, glandular structures are present behind the brain, which show histological signs of secretory activity before the moult and regress in the early adult stages: they consequently could be analogous to the insect thoracic glands[188] (see pp. 88–89). In the King-crabs (Xiphosura) many neurosecretory cells have been described throughout the central nervous system: their density in different regions is correlated with the chromatophorotrophic activity of extracts of these regions.[233]

It is likely that the overall endocrine control of moulting in the arthropods is similar to that in insects and crustaceans (see Chapters 6 and 9).

Since little detailed experimental work has been performed on arthropods other than these two major groups, the remainder will not be discussed further.

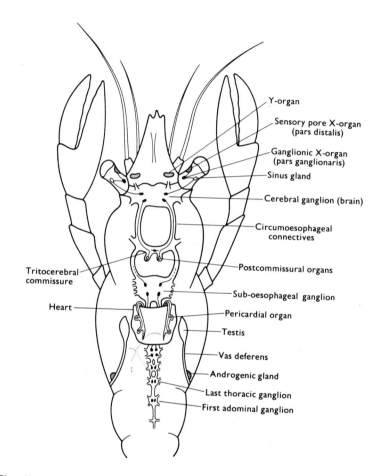

Fig. 2.19 The endocrine system of a generalized male crustacean. Neurosecretory cells (shown in black) are found in the eyestalks, brain, sub-oesophageal ganglion and throughout the remainder of the central nervous system. The neurohaemal organs (shown in white) may be supplied by several groups of neurosecretory cells. The sinus glands receive axons from the ganglionic X-organ and from the brain; the post-commissural organs receive axons from the brain; and the pericardial organs, axons from the thoracic ganglia. Epithelial endocrine glands (cross hatched) are the Y-organs and the androgenic glands. (After Gorbman and Bern[118])

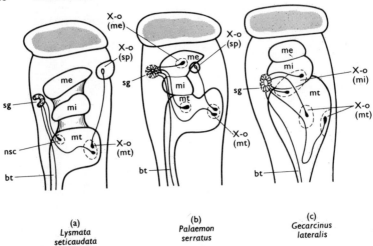

Fig. 2.20 Arrangements of neurosecretory cell groups within the eyestalks of different crustaceans. Dark areas represent neurosecretory cells, light ovals their axon terminals.

(a) *Lysmata seticaudata*. The X-organ consists of neurosecretory cells in the medulla terminalis, with the sinus gland as its major neurohaemal organ. The sinus gland also receives axons from neurosecretory cells in the brain. But some neurosecretory cells in the medulla terminalis send axons to the sensory pore 'X-organ' which, despite its name, is another neurohaemal organ.

(b) *Palaemon serratus*. There are two areas of neurosecretory cells, called ganglionic X-organs, one in the medulla terminalis and one in the medulla externa, both of which send axons to the sinus gland where they are joined by neurosecretory axons from the brain. The sensory pore is absent in this species, but the sensory pore 'X-organ' remains, supplied by neurosecretory axons from the medulla terminalis.

(c) *Gecarcinus lateralis*. In this species there are several ganglionic X-organs in the medulla terminalis and the medulla interna with axons terminating in the sinus gland together with those from the brain.

Abbreviations: bt: neurosecretory axons from the brain terminating in the sinus gland; me, mi, mt: medulla externa, interna and terminalis; nsc: neurosecretory cell group; sg: sinus gland; sp: sensory pore; X-o: X-organ.

(After Gorbman and Bern[118])

THE ORIGIN OF NEUROSECRETORY CELLS

Neurosecretory cells have been defined as neurones which also show the cytological characteristics of gland cells. They receive nervous impulses, but these are not transmitted to other neurones or effector cells: instead the neurosecretory axons terminate in close proximity to parts of the circulatory system, and release substances which act on distant effector organs or upon endocrine glands. These neurosecretory substances, therefore, are themselves hormones.

This definition implies that neurosecretory cells are modified neurones, with a glandular function superimposed upon their fundamental neuronal character.[122] But the opposite could equally well be true: neurosecretory cells may have been originally gland cells which later acquired some of the properties of neurones.[52] The morphological relationships of the nemertean cerebral organ (p. 9) have been used to support this view.[232]

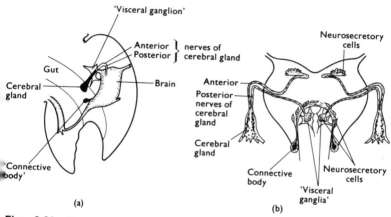

Fig. 2.21 Neurosecretory cells and neurohaemal organs in *Jonespeltis splendidus* (Myriapoda). (a) From the side, (b) in horizontal section. Groups of neurosecretory cells in the brain and visceral ganglia send axons to the major neurohaemal organs, the cerebral glands. Other cells are associated with the connective bodies, which are likely also to be neurohaemal organs. (After Prabhu[225])

A similar incorporation of epidermal mucous cells into the posterior lobes of the brain in the polychaete nephtyids (Fig. 2.22) could be another example.[53] In the phyllopod crustaceans and copepods there is a sensory frontal organ often associated with a large secretory cell. The frontal organ is incorporated into the central nervous system in the malacostracan Crustacea, where it forms an X-organ, again revealing a connection between an originally epidermal glandular (and sensory) structure and a later neurosecretory centre.

But evidence for the origin of neurosecretory cells derived from *living* representatives of modern animal groups must be considered extremely tentative. Indeed, the widespread occurrence of neurosecretory cells in animals as diverse as coelenterates and man, together with the fact that in arthropods and vertebrates the major neurosecretory cell

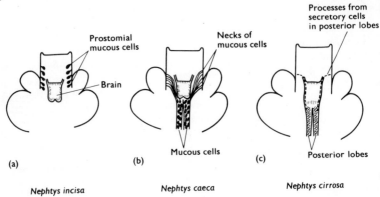

Fig. 2.22   Incorporation of mucous cells into the brain of nephtyids. In (a), the cells are superficial and separate from the brain. In (b), the cells form two lobes posterior to the brain, but their necks open externally. In (c), the posterior lobes of the brain contain neurones and gland cells, although processes still extend from the cells to the epithelium. (After Clark[53])

groups are found in the phylogenetically oldest parts of the brain, suggest a very ancient origin for the cells. Arguments for either a neuronal or a glandular origin of neurosecretory cells become largely irrelevant if it is assumed that the primaeval metazoan nervous system had two major functions, enabling the animal to react quickly to short-term fluctuations in its surroundings and also to co-ordinate developmental and other processes with much more extended environmental change. The neurosecretory cell would thus be as old as the neurone, and speculation about its nervous or glandular origin as unrewarding as that concerning the chicken and the egg.

Much that is known about the function of neurosecretory cells in many animals (Chapter 14) supports this point of view. The morphological similarities between neurosecretory cells and neurones would merely reflect a common origin. Indeed, functionally there is less difference between neurosecretory cells and neurones than is commonly accepted. Neurones are as glandular as neurosecretory cells, manufacturing and secreting chemicals: the essential difference between them is in the kind of chemical produced. Neurones produce materials such as adrenaline and acetylcholine which have a local action at the synapse and are rapidly destroyed. Neurosecretory cells produce a range of peptide or small molecular weight proteins which are comparatively long-lived and act at a distance. But all these substances are secretory products of the cells. In many insects, neurosecretory axons terminate in the corpora allata (Fig. 2.12), and their secretions have a local action upon the cells of the

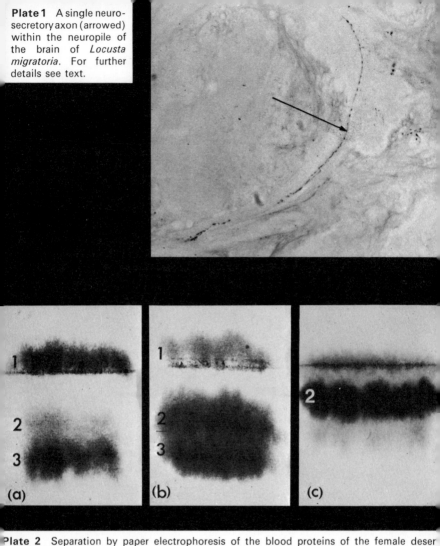

Plate 1 A single neurosecretory axon (arrowed) within the neuropile of the brain of *Locusta migratoria*. For further details see text.

Plate 2 Separation by paper electrophoresis of the blood proteins of the female desert locust, *Schistocerca gregaria*. This is a relatively crude electrophoretic technique, separating proteins only into major groups. Normally, three groups of proteins are separated by paper electrophoresis of locust blood.

(a) The three protein groups in a female before the start of vitellogenesis. (b) Those in a female during vitellogenesis. (c) Those in a female 10 days after total ovariectomy.

The changes in blood protein concentration during vitellogenesis (see Fig. 7.19) are due almost entirely to fluctuations in band 2. Before the start of vitellogenesis, band 2 is faint (a) but increases in density as vitellogenesis proceeds (b). After total ovariectomy, the blood protein concentration increases considerably, reflected in the very much greater density of band 2 (c). (see p. 128).

After prolonged electrophoresis, band 2 may separate into two protein groups; it is therefore likely that at least two vitellogenic proteins are present in the blood, to be taken up eventually by the developing oocytes.

glands.[235, 214] In other insects, it is claimed that *all* the neurosecretory axons terminate within the organs they control[156] (Fig. 2.23). And in many animals, including the vertebrates, some neurosecretory axons terminate in different parts of the neuropile of the central nervous system (Plate 1) perhaps affecting transmission across neuronal synapses.[13] So

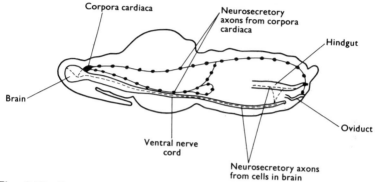

Fig. 2.23 Part of the closed neurosecretory system of the aphid. Neurosecretory axons from the brain and corpora cardiaca run to many parts of the body. Those innervating the hindgut and oviduct are shown here (After Johnson[156]).

the distinction between neurosecretory cells and ordinary neurones becomes even more blurred, not unexpectedly if the two kinds of cell have a similar, and ancient, origin. The implications of considering neurosecretions to be the 'oldest' animal hormones will be discussed in Chapter 14.

3

Endocrine Mechanisms in the Coelenterata, Turbellaria, Nematoda and Echinodermata

In the previous chapter, neurosecretory cells were said to occur within the nervous systems of a number of the less highly organized invertebrates. Neurohaemal organs have not been found in these species, nor have epithelial endocrine glands. The difficulties of ascribing functions to dispersed neurosecretory cells, described in Chapter 1, are encountered immediately in these groups.

COELENTERATA

Growth, regeneration, and the development of sexuality in species of *Hydra* provide the most convincing evidence for the existence of hormones in the Coelenterates.[29, 30, 31] In normal *Hydra pseudoligactis*, the growth region lies just proximal to the hypostome and active cell proliferation occurs here. In this region, many nerve cells are found with long, broad axons which are enlarged at intervals into vesicles which often contain droplets which can be stained with vital dyes[29] (Fig. 2.1). The cells may also be stained in fixed material with paraldehyde-fuchsin —a stain much used to differentiate neurosecretory cells in other animals. But in *Hydra*, paraldehyde-fuchsin also stains nematocytes, mucous droplets in musculo-epithelial cells, and other materials, and vital staining of the nerve cells is to be preferred.

When the hypostome of *Hydra pseudoligactis* is cut and removed from the body column, the numbers of droplets in the nerve cells increase considerably for up to four hours after the operation, and then decrease during the subsequent two hours. But during this latter period, the droplets appear within the tissues surrounding the nerve cells, apparently being released from the nerve cells. There seems every justification for calling these droplet-containing neurones neurosecretory cells. Rapid cell proliferation and regeneration of the body column follows the induced activity of neurosecretory cells in the excised hypostome.[30]

The presence of neurosecretory cells almost exclusively in the growth region of *Hydra*, and their hyperactivity when the accelerated growth characteristic of regeneration is induced, provides strong circumstantial evidence for the production of a growth hormone by the neurosecretory cells. Moreover, normal growth in *Hydra* is almost invariably accompanied by budding. The development of the bud is initially controlled by the parent's growth centre: it dies if prematurely separated. But just as the tentacle outpushings appear on the bud, the bud's own growth centre is developed—and this coincides exactly with the first appearance of neurosecretory cells in this region of the bud.

But there is more direct evidence for the existence of a growth hormone produced in this region. Extracts of the hypostome will stimulate growth, in the form of supernumerary heads, when applied to any part of the body column (Fig. 3.1). Extracts of other parts of the body do not have this

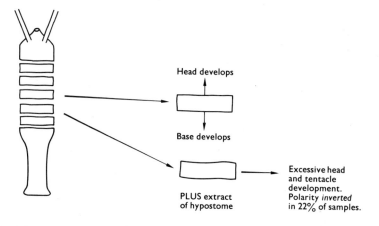

Fig. 3.1 Any part of the column of *Hydra* will reconstitute an individual when isolated. But when hypostomal extract is added, many supernumerary heads and tentacles are formed and in a proportion of samples these develop on the base side of the sections. (After Burnett, Diehl and Diehl[30])

effect. It must be emphasized that it is impossible to be sure that the hypostomal neurosecretory cells are the exact source of the active factor in the extracts, but such an assumption is supported strongly by the observations described above.

Neurosecretory cells in the sub-hypostomal growth region of *Hydra pseudoligactis* are scarce or absent in fully sexual forms. Hypostomal extracts from sexual *Hydra pirardi* will not induce growth when applied to the body columns of other individuals. It seems likely, therefore, that the disappearance of the neurosecretory cells, with the consequent absence of growth hormone, is associated with sexual development.[31]

When the hypostome of a growing *Hydra pirardi* is grafted onto the body column of a sexual *Hydra fusca*, the testes of the latter do not mature. Instead, they begin to develop into small buds (Fig. 3.2). The growth hormone therefore inhibits the onset of sexuality.

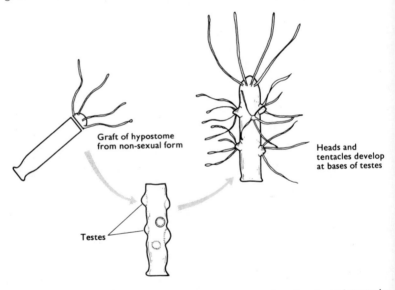

Graft of hypostome from non-sexual form

Heads and tentacles develop at bases of testes

Testes

Fig. 3.2 A sexual *Hydra fusca* transforms its testes into heads and tentacles when its hypostome is replaced with that from a non-sexual individual. The non-sexual hypostome contains a factor which favours somatic rather than reproductive development.

In *Hydra*, the interstitial cells can develop into nematocytes, nerve cells, or other somatic structures. But they can also differentiate into gametes, and in species in which there is no large store of interstitial cells, somatic and reproductive differentiation are antagonistic. Thus the tentacles of the sexual forms of *Hydra pirardi* can be quite devoid of nematocysts.

When individuals of *Hydra pirardi* with well formed testes are severed through the middle of the body column, they will remain in the sexual state for a considerable time. But when the sub-hypostomal regions of growing individuals are grafted onto the excised surfaces of the sexual forms, then the cells at the bases of the gonads show extensive differentiation into nematocytes within 24 hours. As long as the interstitial cells have not undergone their first meiotic division, the presence of growth hormone switches their development from gamete formation to nematocyte development (Fig. 3.3), in other words away from reproductive and

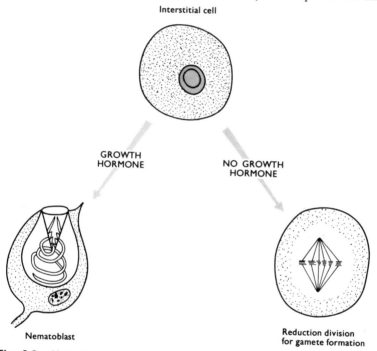

Fig. 3.3 Alternative developmental pathways for an interstitial cell in *Hydra*. In the presence of growth hormone from the hypostomal region, the cell transforms to a nematoblast; in the absence of the hormone, meiosis occurs, leading to gamete formation.

towards somatic development. Very often, the grafted individuals will develop supernumerary heads or tentacles.[30]

The neurosecretory growth hormone produced in the sub-hypostomal region therefore has two functions: it induces cell proliferation and growth, and it directs interstitial cell development into a somatic course. In the absence of the growth hormone, growth naturally ceases, but in addition, the interstitial cells then develop along a reproductive course.

It follows from this that growth and reproduction in *Hydra* are antagonistic processes, as they are in many other animals. In different species of *Hydra*, various environmental conditions will inhibit growth, and consequently stimulate sexuality. In *Hydra pseudoligactis*, bringing animals from a temperature of 20–21°C to 8°C will induce egg formation within 30 days. In *Hydra pirardi* the same treatment causes testes formation. But reducing the temperature inhibits sexuality in *Hydra littoralis*. In *Hydra pseudoligactis* and *Hydra pirardi*, a progressive shortening of the day length will induce sexuality, although the effect must in some way be combined with temperature because increasing daylength during the Spring will also induce sexuality in *Hydra pseudoligactis*.[31]

Besides temperature and photoperiod changes, sexuality in *Hydra* may be induced by starvation, an increased carbon dioxide concentration, stagnation in the culture medium, and so on. All these factors are likely to interfere with growth, either directly, or indirectly through an effect upon the sub-hypostomal neurosecretory cells. The resulting development of sexual forms, with the eventual production of a resting zygote, is clearly of advantage to the species in overcoming the unfavourable conditions. In some species the reaction to changes in daylength also favours the production of an overwintering zygote. It is of considerable interest that neurosecretory mechanisms are involved in these protective adaptations in Coelenterates in much the same way as they are in much more complicated animals like the insects (see p. 142 *ff*).

TURBELLARIA

Freshwater planarians, such as *Euplanaria* (*Dugesia*) *lugubris* and *Polycelis nigra*, have for long been known to possess quite extraordinary powers of regeneration. Undifferentiated cells, the neoblasts, migrate to the vicinity of a wound or an amputation, and are then able to divide and to differentiate into any tissue which has been damaged or removed. The neoblasts are thus both motile and totipotent. If the wound which they repair is extensive, they form a blastema of regenerating tissue rather similar to that found in polychaete worms after segment amputation (p. 46).

What initiates the migration of neoblasts towards the damaged tissues? In *Euplanaria lugubris*, x-irradiation of the anterior part of the body at a level sufficient to destroy most of the tissues is followed by necrosis of the irradiated region and the subsequent death of the individual.[189] But if part of the irradiated region is cut off just before or after x-irradiation, then neoblasts migrate from the posterior, non-irradiated part of the body, and replace not only the amputated section, but the rest of the irradiated region also (Fig. 3.4). The result is the same when the cut is

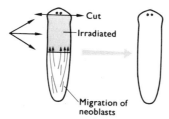

Fig. 3.4   Regeneration in *Euplanaria lugubris*. When an anterior cut is made and part of the body subsequently irradiated, the whole of the irradiated and amputated regions are regenerated from neoblasts which migrate from the posterior parts of the body. Compare Fig. 3.5.

made up to four days before subsequent irradiation; but a longer interval between section and irradiation prevents regeneration. If the cut is made some distance from the anterior border of the individual, and the area is then irradiated, regeneration will not extend anterior to the cut (Fig. 3.5). This region becomes necrotic and dies.

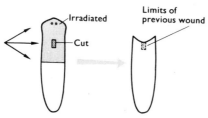

Fig. 3.5   Regeneration in *Euplanaria lugubris*. When a wound is made behind the anterior tip of the body, and the region subsequently irradiated, regeneration does not occur anterior to the wound. The wound is the source of factors (wound hormones) which stimulate the neoblasts to migrate and regenerate the damaged tissues.

The results of these experiments on *Euplanaria lugubris* suggest very strongly that the neoblasts are activated by, and migrate towards, some factor produced by the damaged region of the body. The first steps in regeneration are thus set in motion by, for want of a more precise term, 'wound hormones' produced by the damaged cells and tissues. Little is known about these substances: but their presence in wounded tissues has been inferred not only in planarians, but in worms, insects and vertebrates too.

Having arrived at the site of damage, what controls the differentiation of neoblasts into specific cells and tissues? This problem has been examined in the control of eye regeneration in *Polycelis nigra*. This planarian, as its name implies, has many (about 80–90) eyes on the anterior borders of the head. If an eye-bearing region is amputated, ocular regeneration occurs in about 7 days at 18°C. But if the brain is removed at the same time, the eyes do not regenerate at all. This absence of regeneration is not due to removal of the ocular nerves, because the retinal and capsule cells begin regeneration, in normal animals, *before* the nerves are formed. Moreover, a section of the anterior border with eyes removed from one individual will regenerate eyes when grafted onto a normal host, in the vicinity of the brain, without any nervous connections between host and graft being established (Fig. 3.6). Thus the brain is

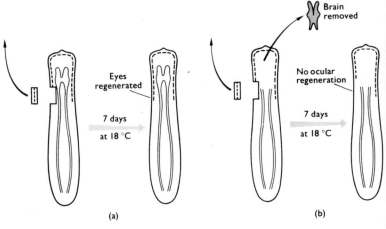

Fig. 3.6 Ocular regeneration in *Polycelis nigra*. In the presence of the brain, eyes are regenerated when part of the ocular region is amputated (a). Where the brain is removed simultaneously with amputation of part of the ocular region, eyes are not regenerated (b). The brain is necessary for ocular regeneration in *Polycelis*.

necessary for ocular regeneration in *Polycelis*, and it exerts its effect at a distance: it would seem that the brain produces a regeneration hormone.[189]

Neurosecretory cells are present in the posterior ventral parts of the brain of *Polycelis*.[189] Homogenates of the anterior parts of planarians added to the water in which they are reared, will induce ocular regeneration in decerebrate individuals. The active principle will withstand temperatures of 60°C for 10 minutes, but is destroyed by boiling for 30 minutes. A homogenate of 5 heads in 10 ml of water produces only a little ocular regeneration; 10–15 heads in the same volume of water produce normal regeneration; 20 heads accelerate regeneration by 24–48 hours.

When an anterior segment, with eyes removed, is grafted to the posterior part of a normal animal, the eyes do not regenerate[189] (Fig. 3.7). Clearly, the brain cannot influence caudal tissues. But when tails

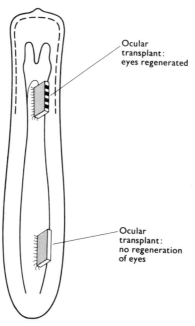

Ocular transplant: eyes regenerated

Ocular transplant: no regeneration of eyes

Fig. 3.7 Ocular regeneration in *Polycelis nigra*. Part of the ocular region transplanted from another individual to a site near to the host brain will regenerate eyes. Transplanted to a posterior site, no eyes are regenerated. The influence of the brain in inducing regeneration does not extend very far posteriorly (Figs. 3.4–3.7 after Dubois and Lender[77a]).

are homogenized, and the extract added to water in which decerebrate *Polycelis* with eye segments removed are kept, some ocular regeneration will occur. If the tails are first treated with 70% alcohol, or heated to 60°C for two minutes, or are finely ground with sand, then the extracts are more potent in inducing ocular regeneration. It is concluded that not only is there a gradient of regeneration hormone, high anteriorly and decreasing caudally, but that in addition the cells in the tail region actually absorb or sequester the hormone to reduce its effectiveness in the normal animal.[189]

The regeneration hormone cannot be definitely identified as a product of the neurosecretory cells in the brain. Moreover, although the hormone is necessary for ocular regeneration, eyes are only formed in those regions where they normally occur. The interaction of some other factor with the regeneration hormone must determine that neoblasts in this region will differentiate into eyes. The situation is thus rather similar to the control of segment regeneration in polychaete worms (p. 52). It must also be emphasized that this property of evoking regeneration is not confined to the *Polycelis* brain. Other substances, particularly extracts of chick embryos, will also induce ocular regeneration. The evidence for the existence of neurosecretory hormones in flatworms must therefore be considered rather slender.

NEMATODA

Nematodes moult their cuticles in a manner very similar to that of the arthropods (p. 78). A new cuticle is first deposited under the old, and the latter is then partly digested by moulting fluid before splitting along a circumferential line of weakness to allow the emergence of the next developmental stage. One of the enzymes responsible for digestion of the old cuticle is leucine aminopeptidase.[74]

Phocanema decipiens is parasitic in the muscles of the cod, where it occurs as a last stage larva, and in the digestive tract of the seal, where the final moult to the adult stage takes place very soon after ingestion. This final moult can be duplicated in a suitable incubation medium *in vitro*, enabling accurate observations to be made on the control of the moult.

When fourth stage larvae of *Phocanema* are removed from the cod and placed in the incubation medium, a new cuticle begins to be formed within 12 hours, and ecdysis (the moulting of the old cuticle) occurs 3 to 5 days later. Two groups of neurosecretory cells are present in *Phocanema*, one in the dorsal and the other in the ventral ganglion (p. 12). Within 12 hours of being placed in the incubation medium, neurosecretory granules appear in the cells, reaching a peak of accumulation between the second and fifth days. After the sixth day, the granules have disappeared.[74]

Although these histological changes suggest that the neurosecretory cells are active between the second and fifth days of incubation, there is no evidence that neurosecretion is released from the cells during this time. In fact, when the larvae are ligatured to prevent any hormone, if present, from reaching the posterior parts of the body, new cuticle is deposited normally both in front of and behind the ligature. The activity of the neurosecretory cells is not concerned with cuticle deposition.[74]

Leucine aminopeptidase is secreted by the so-called excretory gland. In *Phocanema*, a cycle of synthesis and release of the enzyme occurs at the time of ecdysis. If slight changes are made in the constitution of the incubation medium, the neurosecretory cells do not show their cycle of activity, there is no leucine aminopeptidase production in the excretory gland, and ecdysis is inhibited, although new cuticle formation proceeds normally. Are the neurosecretory cells therefore concerned, not with cuticle deposition, but with ecdysis?

When excretory glands are incubated with saline extracts of the anterior ends of *Phocanema* in which the neurosecretory cells are empty, there is no increase in leucine aminopeptidase in the glands. But incubated with similar extracts of *Phocanema* with full neurosecretory cells, the leucine aminopeptidase activity of the excretory glands increases considerably.[74] It is likely, therefore, that in normal *Phocanema* a hormone from the neurosecretory cells controls the production of moulting fluid enzymes necessary for the casting of the fourth stage larval cuticle. Moreover, conditions must be suitable before the cells are activated to produce their hormone. Presumably the correct conditions are found in the digestive tract of the seal, so that the adult worm eventually emerges from the fourth stage cuticle when its cod host is eaten.

ECHINODERMATA: THE CONTROL OF SPAWNING IN STARFISH

The gonads of the starfish are paired within each of the five arms, lying freely in the perivisceral cavity and opening to the exterior by paired gonoducts, the gonopores situated between the junctions of the arms. When ripe, the gametes are shed into the sea: gametes are shed from all the gonads simultaneously, which suggests some regulating mechanism.

When an aqueous extract of the radial nerves is injected into the perivisceral cavity of one arm, *all* the gonads shed gametes about 30 minutes later. The extracts are not sex specific: radial nerve extracts from males will induce the shedding of eggs, and extracts from females induce the shedding of sperm.[44] Extracts of tube feet, digestive tract, hepatic caecae and gonads are without effect.[44]

The results of these experiments suggest that the control of spawning is hormonal, and since radial nerve extracts alone are effective, that a neurosecretory hormone is involved. When successive layers of the radial nerves are stripped away, either mechanically or by enzymes, only the most ventral part of each nerve is found to contain shedding activity.[44] In starfish with ripe gonads, the ventral layer of the radial nerves contain granules staining with paraldehyde-fuchsin and chrome-haematoxylin-phloxine—which stain neurosecretory material in many other animals. The neurosecretory cells which must be the source of this material in the radial nerves have not yet been identified. But the combination of physiological and histological evidence leaves little doubt about the neurosecretory nature of the hormone.

Isolated ovaries respond to radial nerve extracts by shedding their eggs in the same way as ovaries *in situ*. The hormone must therefore normally act directly upon the gonads, and not through any other endocrine mechanism. It might therefore be suspected that the hormone passes aborally from the ventral regions of the radial nerves, eventually to reach the perivisceral cavities to act upon the gonads. But in fact there is no evidence that such movement occurs. Instead, the hormone is apparently passed out from the nerve into the sea, because a spawning

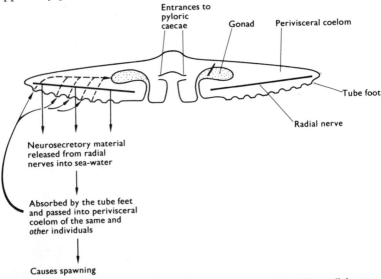

Fig. 3.8 The control of shedding gametes in the starfish. The radial nerves release a neurosecretory substance into the sea water which is taken up by the tube feet into the perivisceral coelom of the same and other individuals. The neurosecretion acts directly upon the gonads and causes shedding of the gametes. The substance can thus be considered as both a hormone and a pheromone, since it affects both the individual that produces it, and other members of the same species.

starfish will stimulate others in its vicinity to spawn[44] (Fig. 3.8). The tube feet of the starfish can take up substances such as amino acids from the surrounding medium, which appear in the perivisceral cavity within two minutes. The shedding hormone is a heat labile, dialyzable polypeptide, containing 10–15 amino-acids and with a molecular weight of about 2600.[44] It would thus seem capable of being taken up by the tube feet. If this mechanism does in fact operate, since neighbouring individuals can react to its presence in the sea, the shedding substance acts both as a hormone (for the individual which produces it) and as a pheromone (p. 231) for its effect upon other starfish.

In the female starfish, the hormone acts by dissolving the cementing substance by which the oocytes adhere to the follicle cells and to the ovarian wall.[44] This is followed by maturation divisions in the oocytes. It is likely that these divisions are caused by a second hormone produced by the ovaries. This hormone, of lower molecular weight than the shedding hormone, has been separated by gel filtrations.[44] It also has some shedding activity when assayed on ripe gonads, so it could supplement the action of the neurosecretory hormone.

High concentrations of radial nerve extracts can *inhibit* the shedding of gametes. This suggests that the radial nerves also contain an inhibitory hormone, with a much higher concentration threshold than the shedding hormone. The two hormones in radial nerve extracts can be separated on Sephadex columns. When radial nerves are assayed throughout the year, the inhibiting hormone is found only in animals with ripe gonads, while the shedding hormone is present continuously.[44] This apparent contradiction is explained by assuming that gamete shedding is primarily controlled by fluctuation in the concentration of the inhibiting hormone. Premature spawning is most likely as the gonads approach maturity: the high level of inhibitory hormone would prevent spawning during this dangerous period. When the concentration of inhibiting hormone begins to decline, the shedding hormone becomes increasingly effective. It is likely that a hormone inhibiting gamete shedding becomes imperative when the shedding hormone itself has pheromonal properties (p. 231).

These experimental studies on growth, regeneration and reproduction in a very limited number of 'lower' invertebrates provide evidence, albeit fragmentary and circumstantial, for the intervention of neurosecretory hormones in the control of these processes. In *Hydra* and *Phocanema* especially, there is also evidence for the view that the neurosecretory mechanism is itself influenced by environmental factors. In the next chapter, the endocrine control of the same processes of growth, regeneration and reproduction in the annelids will be discussed. In this group the neurosecretory control over these processes shows a startling similarity to that already described for *Hydra*.

4

Endocrine Mechanisms in the Annelida

Like the species already discussed no annelid possesses epithelial endocrine glands. Any hormonal control over development is therefore likely to be due to neurosecretions. In 1936, neurosecretory cells were first demonstrated in the supra-oesophageal ganglion, or brain, of a polychaete.[231] But twenty years were to elapse before it was shown that growth, regeneration, asexual reproduction and maturation of the gonads together with epitokal metamorphosis (when present) are all controlled by hormones from the brain. In the oligochaetes and leeches also, neurosecretory cells are present in the brain,[100, 127, 120] although their functions are less well defined than those of the polychaetes.

GROWTH AND REGENERATION IN POLYCHAETES

The typical annelid body consists of a presegmental prostomium and a postsegmental pygidium separated by a variable number of metameric segments (Fig. 4.1). Oligochaetes and leeches hatch from the egg as segmented individuals; the polychaetes become segmented only after metamorphosis from the unsegmented trochophore larva. But in all annelids, growth consists of a proliferation of segments from an actively growing zone in front of the pygidium, followed by enlargement of the new segments. Where segment proliferation continues throughout the life of the annelid, as in many polychaetes, its rate decreases considerably as the individual ages.

The soft-bodied annelid is very vulnerable to attack by predators or

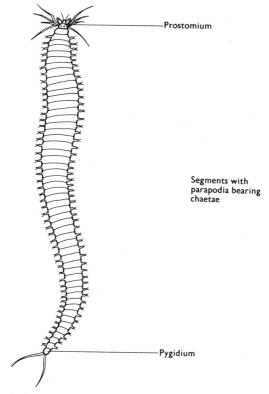

—Prostomium

Segments with
parapodia bearing
chaetae

—Pygidium

Fig. 4.1   Dorsal view of a polychaete annelid. The body consists of the pre-segmental prostomium behind which are a variable number of metameric segments and, at the posterior end, the post-segmental pygidium. During growth, new segments are formed in the proliferative zone just in front of the pygidium.

damage by physical agents. An ability to regenerate lost parts of the body, including specialized sensory or reproductive organs, is clearly of great advantage to the individual. In annelids, caudal regeneration is wide-spread: the posterior parts of the body are more likely to be lost under natural conditions. Nereid polychaetes readily regenerate a new tail, but only rarely a new head. Sabellid and syllid polychaetes, some earthworms and leeches, will regenerate new heads; and the highly specialized tube-dwelling *Chaetopterus* will regenerate a complete individual from any one of its first fourteen segments.

When segments are lost, the open wound is sealed by the contraction of the body wall muscles and then a blastema is formed, composed of mesoderm with overlying ectoderm, together with neoblast cells of mesenchymal origin (Fig. 4.2). The blastema produces the ectodermal and mesodermal parts of the regenerating segments, and proliferating endoderm forms the gut. Once a new pygidium has been formed, the blastema takes over the role of the original proliferative region.[129]

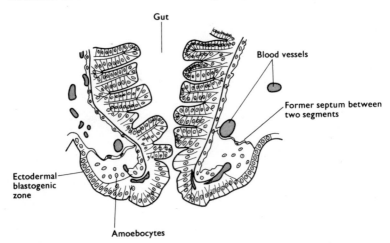

Fig. 4.2 Diagram of a median frontal section through the healing zone of *Nereis diversicolor* eleven days after amputation of posterior segments. The wound has healed by the abutment of the (ectodermal) body wall and the (endodermal) gut wall. The gut remains open. The blastema is beginning to be formed by ectoderm cells (ectodermal blastogenic zone). These cells, and later the mesodermal cells in the area, de-differentiate and form the new region of segment proliferation. Some epithelial cells are histolysed during this process: amoebocytes may aid this breakdown. (After Herlant-Meewis[129])

Growth is stimulated by both local and systemic influences. It is likely that in the annelids **wound hormones** (unspecified chemicals released from damaged cells, see p. 37) play an important part in blastema formation through stimulating the activation and migration of neoblasts.[129] Moreover, the nerve cord exerts a local trophic effect upon regeneration: not only is its presence necessary, but abnormal regeneration will occur if it is deflected from its usual position (Fig. 4.3). In this way, regeneration can be induced along the whole length of the body in *Myxicola aesthetica*. The influence of the nerve cord is non-specific: it induces regeneration, but the *kind* of regeneration which occurs varies according to the region of the body.[129]

Equally important for regeneration is a systemic or hormonal influence. Little is known at present of any possible hormonal regulation of cephalic regeneration. The polychaete *Lycastrus* does not normally

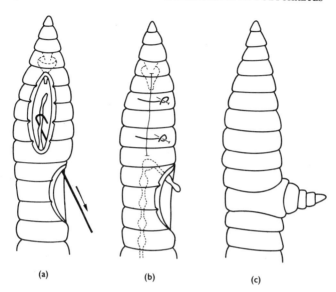

(a)　　　　　　　(b)　　　　　　　(c)

Fig. 4.3 The induction of a lateral head by deflection of the nerve cord in an oligochaete. (a) and (b) show the technique of cutting and deflecting the nerve cord. Such deflection in this cephalic region causes the regeneration of a small supernumerary head (c). If the anterior nerve cord is deflected even further posteriorly, an undifferentiated protruding bud results and not a supernumerary head. Thus although deflection of the nerve cord results in abnormal regeneration, the *kind* of regeneration depends upon the region of the body into which the nerve cord is deflected. (After Herlant-Meewis[129])

regenerate a head after decapitation, but will do so if the headless body is given a brain—which is unlikely to happen in nature.

The hormonal regulation of growth and caudal regeneration is well established. The ragworm *Nereis diversicolor* provides most of the information. Normally, the young worm can regenerate posterior segments. If, however, its brain is removed, growth stops, and amputation of the posterior segments from a decerebrate individual is not followed by regeneration[42, 79, 57] (Fig. 4.4). Subsequent implantation of a brain from a growing worm restores both growth and the power of regeneration.[58] Caudal amputation from a fully grown *Nereis* is not followed by regeneration, even in the presence of the brain.

Proliferation of new segments is the essential feature of both growth and regeneration. The only distinction between the two processes is that segment proliferation is initially very rapid following amputation. Even so, as regeneration proceeds the rapid rate of segment proliferation declines to that of normal growth and it is impossible to determine when

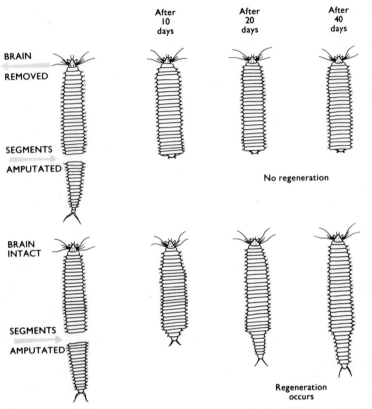

Fig. 4.4   Effect upon caudal regeneration of removing the brain from a growing *Nereis diversicolor*. Amputated posterior segments are regenerated as long as the brain remains *in situ*; when amputation is accompanied by brain removal, segment regeneration is prevented.

regeneration finishes and normal growth recommences[114] (Fig. 4.5). So apart from the initial regeneration of the pygidium, growth and regeneration must be essentially similar, but occurring at different rates.

In *Nereis*, the presence of the brain is necessary for caudal regeneration. But what determines the number of segments to be regenerated? In practice, it is found that the number of segments regenerated more or less equals the number amputated (Fig. 4.6). This is so when the worm's own brain is left in place, or when it is replaced by a brain from another individual.[114] How does the *Nereis* 'know' the number of its segments which have been amputated, particularly when it has been given a new brain, unconnected with any part of its own central nervous system?

One way would be for the brain to produce *more* hormone as more segments are amputated. In other words, there could be a feedback from the region of amputation to the brain, and the rate of segment proliferation would then be directly proportional to the amount of hormone produced.

Until recently, it was thought that segment regeneration in *Nereis* was controlled in just such a manner. The brain was activated to manufacture

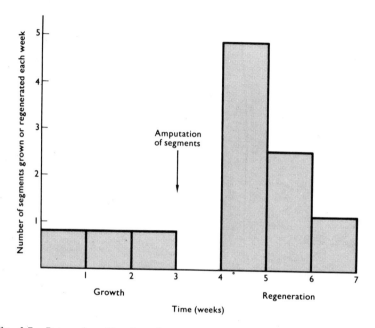

Fig. 4.5   Rates of proliferation of new segments in *Nereis diversicolor* during normal growth and after caudal amputation. After caudal amputation, the rate of segment proliferation increases and then gradually decreases as regeneration proceeds. Thus segment proliferation during normal growth and during regeneration is essentially the same process but occurring at different rates. (After Golding[114])

hormone by the amputation of tail segments, and the hormone accumulated in the brain for the subsequent 3 or 4 days. Then at a critical period, the large amount of hormone was released into the blood, inducing a degree of regeneration proportional to its concentration.[245]

But this hypothesis cannot be valid as the following experiment shows. A brain is taken from a young, intact worm and implanted into a decerebrate, tail-less individual. One day later it is removed, and replaced by

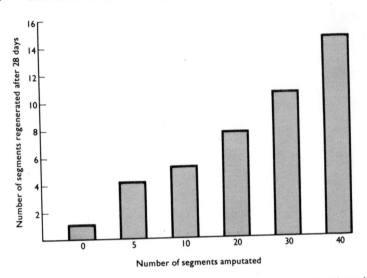

Fig. 4.6 The relationship between number of segments amputated and number regenerated in *Nereis diversicolor*. The worm eventually regenerates a number of segments approximately equal to the number amputated. (After Golding[114])

another similar brain. This is left for one day and replaced, and so on for 10 days. Segment regeneration occurs in all such host individuals, the amount of regeneration being related to the number of amputated segments. But clearly the amount is not dependent upon variable amounts of hormone accumulating in the brain, because *all* animals are given similar numbers of brains for the same short periods of time.[114]

It is possible, of course, that if the implanted brains all secrete hormone at the same rate, the concentration of the hormone in a *Nereis* with many segments amputated would be greater than in one with only a few segments amputated, simply because the volume of body in the first instance would be considerably less than in the second. So even without feedback from the amputated region, the blastema might react to different hormone concentrations. But this possibility is dismissed by the results of the next experiments.

The posterior half of one *Nereis* is grafted on to the body of another and different numbers of segments are amputated from each. Both the host and the graft regenerate segments appropriate to the number lost (Fig. 4.7), although both are dependent upon the host's brain, and must be reacting to the presence of the same concentration of hormone.[11] Moreover, if an already regenerating *Nereis* is grafted on to a host, which then has segments amputated, the graft *continues* to regenerate at the

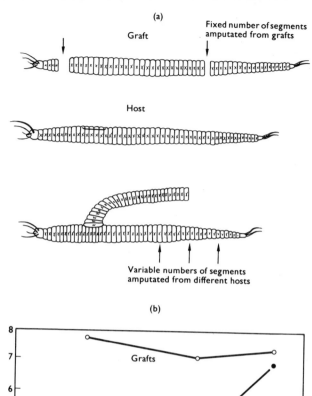

(a)

Graft

Fixed number of segments amputated from grafts

Host

Variable numbers of segments amputated from different hosts

(b)

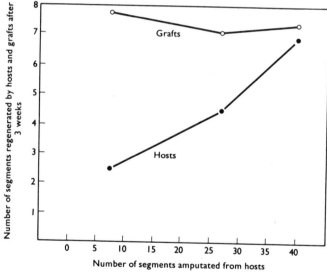

Fig. 4.7 An experiment to demonstrate that the number of segments regenerated is independent of the concentration of brain hormone in the blood of *Nereis diversicolor*. (a) The posterior part of one worm is grafted onto an intact (host) worm. When segments are amputated from both host and graft, regeneration in each will depend upon hormone derived from the host brain: the concentration of brain hormone must be the same in host and graft. (b) The number of segments amputated from the hosts is varied, but the number of segments amputated from the grafts is always the same: although regeneration in the hosts therefore varies, regeneration in the grafts is seen always to remain the same. Consequently, host regeneration cannot be the result of differing hormone concentrations, since the graft is unaffected. (After Golding[115])

rate it had reached when it was joined to the host, and the host regenerates at its usual rate.[115] Thus both the rate of regeneration and the number of segments produced can be very different in an identical hormone environment.[115]

In *Nereis*, it is certain that the brain is necessary for regeneration, and probably for normal growth, but its hormone does not act in any regulating capacity. Instead, the control of segment proliferation must be a local mechanism, the property of the tissues at the wound. The brain secretes hormone at a steady rate, but segment proliferation depends upon a changing competence of the blastema to react to its presence. When more segments have been amputated, the blastema is more reactive to the same level of brain hormone.

In arthropod endocrinology, as in vertebrate, the principle is well established that fluctuating levels of hormones bring about differential changes in their target organs. It is often forgotten that the state of the target itself can modify the apparent effects of hormones. In *Rhodnius prolixus*, for example, epidermal cell division is initiated by developmental hormones (p. 90), but the amount of division is affected by the degree of abdominal stretching brought about by the blood meal, and by other less obvious factors such as the position of epidermal cells in the body. Imaginal buds in holometabolous insects may not metamorphose until they have reached a certain degree of development. In *Nereis*, the overriding importance of the target in reacting differentially to the same hormone concentration is particularly well demonstrated.

What determines this differential reactivity of the blastema after amputation of various numbers of segments? One favoured answer is that there is an axial gradient of growth potential in the intact animal, high anteriorly and declining posteriorly. But this explanation really only restates the problem in a different way, for at present the nature of this axial gradient is obscure. Such gradients of influence have been recognized in the embryonic development of many animals, including vertebrates, and in the regeneration of coelenterates and flatworms (pp. 32, 39). The hormonal control of growth and regeneration in polychaetes thus abuts onto major problems in developmental biology.

REPRODUCTION AND EPITOKAL METAMORPHOSIS IN POLYCHAETES

Somatic growth makes considerable demands upon the resources available to an animal. The metabolic needs of gametogenesis, particularly in the female, may also be large. Consequently, somatic growth and reproduction are usually antagonistic, occurring alternately as in many crustaceans (p. 191), or separated almost completely, the reproductive period not beginning until somatic growth has ended. The polychaete worms, like the majority of animals, have almost all adopted this second method.

In the polychaetes, preparation for reproduction begins early in life, although full gametocyte development does not start until after the end of somatic growth. Thus *Perinereis cultrifera* begins to accumulate gametocytes in the coelom from early in its second year, although it does not breed until it is three years old. *Nereis diversicolor*, which probably does not live for much longer than a year, has small gametocytes in the coelom by the thirty segment stage, that is, only a few weeks after metamorphosis from trochophore to young worm.

In many polychaetes, the form of the body changes at the approach of sexual maturity. This change in form may be very slight, but it can be so extensive that a highly specialized reproductive individual is produced. In nereid species, such an individual is called a heteronereis and has a greatly modified reproductive region, called the epitoke, with specialized muscles, those of the pre-reproductive period breaking down. It develops membranous frills on its parapodia, and special oar-shaped chaetae; its eyes increase considerably in size, and are much more sensitive to light (Fig. 4.8). These physical changes are associated with behavioural changes: the heteronereis is a more actively swimming form than the pre-reproductive individual.[55] In the syllids, the gonads are confined to the posterior segments, and these segments may break off as a free-swimming unit, often with a newly developed head, but lacking jaws and pharynx. This process of stolonization in the syllids is akin to the asexual reproduction of some oligochaetes, but linked with sexuality and swarming it also bears some resemblance to the formation of the nereid epitoke (Fig. 4.9). There are thus two processes to consider in the control of reproduction in polychaetes: gametogenesis and epitokal metamorphosis.

Endocrine control of gametogenesis

Most of the work on the endocrine control of gametogenesis is concerned with female polychaetes, mainly because oocyte development is easily followed by measurements of cell diameters. In most polychaetes, the oocytes are shed into the coelom at an early stage, and there complete their growth and maturation. More rarely, the oocytes remain in contact with the germinal epithelium, forming discrete gonads attached to the peritoneal walls, from which the eggs are released only during the final stages of maturation. Occasionally, as in *Tomopteris*, abortive oocytes may form nurse cells, but it is more usual for coelomic cells to perform this function.

In *Nereis diversicolor*, the oocytes are shed into the coelom when they

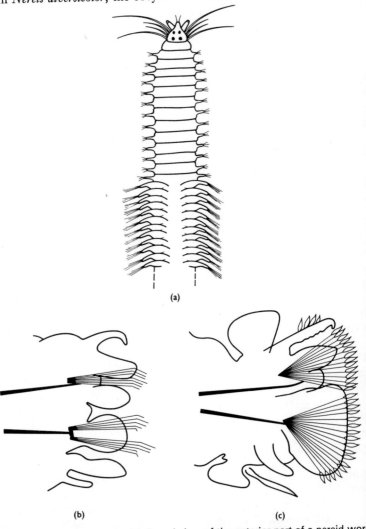

(a)

(b)   (c)

Fig. 4.8   The heteronereis. (a) Dorsal view of the anterior part of a nereid worm showing the structure of the parapodia in the heteronereis form (compare Fig 4.1). (b) Cross-section of parapodium of a pre-reproductive worm. (c) Cross section of parapodium in the reproductive worm. Note the greater number of oar shaped chaetae and increased area of the heteronereis parapodium.

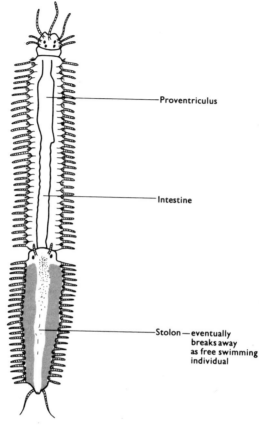

Fig. 4.9 Formation of a stolon in a syllid polychaete. The posterior half of the worm, which contains the gonads, becomes modified and breaks away to form a free-swimming individual.

are 15–20 μ in diameter. Their subsequent growth, to about 70 μ diameter, is slow, but there is then a rapid increase in diameter to about 140 μ, followed once more by a slow growth to a final diameter of 200 μ[56] (Fig. 4.10).

Removal of the brain has no effect upon the proliferation of the oocytes from the germinal epithelium nor upon their initial slow growth. But once they have reached about 30 μ in diameter, brain removal precipitates the rapid growth which normally begins when the oocytes have reached a diameter of 70 μ.[56] Even when this rapid growth has begun normally, brain removal accelerates the process. But the eggs produced

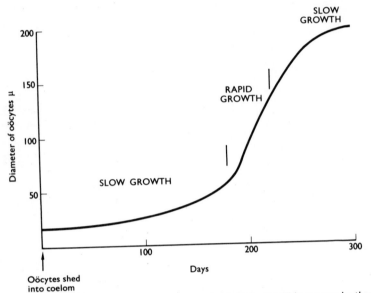

Fig. 4.10 Growth of the oocytes in *Nereis diversicolor*. This occurs in three phases: (1) a period of slow growth from the time the oocytes are shed into the coelom until they reach a diameter of 70 μ; (2) a phase of rapid growth from a diameter of 70 μ to about 140 μ; (3) a further period of slow growth from 140 μ in diameter to the final diameter of about 200 μ. (After Clark[56])

by this precocious and accelerated growth are smaller than usual, and lack their complement of protein yolk platelets. Such eggs can be fertilized, but fail to develop beyond the 8- or 16-cell stage. If the brain is removed from females in which the oocytes have completed their rapid growth phase, the operation is without effect.[56]

When a brain from an immature *Nereis* is implanted into a decerebrate individual, precocious oocyte growth is prevented. Moreover, oocyte development is also inhibited if a brain from an immature individual is implanted into an otherwise normal *Nereis*. Brains from mature individuals, however, have no effect when implanted into either decerebrate or normal worms.

In *Nereis*, therefore, oocyte development is normally inhibited by a hormone or hormones produced by the brain, and vitellogenesis proceeds rapidly when this inhibition is lifted (Fig. 4.11). But this control of oocyte development is not a simple matter: when the inhibition is suddenly removed (by brain removal, for example) although the oocytes grow rapidly, they do not lay down protein yolk normally. Does the brain, therefore, also secrete a hormone that promotes vitellogenesis?

Hauenschild has shown that the brain does not secrete a vitellogenic hormone, at least in the sense of a hormone distinct from the oocyte inhibiting hormone.[125] When the head from a young adult *Platynereis dumerilii* is implanted into a single parapodium of a headless individual, oocyte development in that parapodium is completely inhibited. But implantation of a similar head into a headless fragment of 30 segments is followed by oocyte growth *and* vitellogenesis, whereas in the headless segments without any implanted brain, abnormal vitellogenesis accompanies the rapid oocyte growth. Consequently it is concluded that when

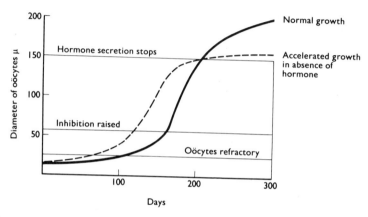

Fig. 4.11 The hormonal control of oocyte development in nereids. A decreasing concentration of a single hormone controls all phases of oocyte development. Initially, a high concentration of the hormone inhibits oocyte development. As the hormone concentration falls, inhibition is lifted and the oocytes grow. Complete absence of the hormone, achieved by removal of the brain, causes accelerated but abnormal growth of the oocytes. Normal progressive growth of the oocytes depends upon a progressive decrease in the amount of hormone. (After Clark[56])

the hormone concentration is high, as in a single parapodium containing an implanted brain, oocyte development is inhibited completely; when the hormone is absent, as in decerebrate individuals, oocyte growth is rapid but vitellogenesis is abnormal; and at an intermediate hormone concentration, obtained by diluting the hormone by implanting a small brain into the headless fragments of 30 segments, oocyte growth *and* vitellogenesis occur. The reduced concentration of inhibitory hormone thus has a *positive* effect upon vitellogenesis. In nereids, therefore, normal oocyte development results from the *progressive* decrease in the amount of a single hormone produced by the brain, and not by the action of two antagonistic hormones[125] (Fig. 4.11).

The situation is rather different in the Arenicolidae (lugworms), for here the brain has a predominantly stimulatory action upon reproduction. In *Arenicola marina* oogonial division and the prophase of the first maturation division take place in the ovary. Nuclear activity is then arrested while the oocytes are released into the coelom and vitellogenesis takes place. Meiotic activity is resumed when spawning is about to occur. Removal of the brain from an immature *Arenicola* has no effect upon vitellogenesis. But brain removal from a mature female delays spawning indefinitely, and the oocytes remain in the suspended prophase of the first maturation division. Spawning in such decerebrate, gravid females is induced by the injection of homogenates of the brain.[151]

The control of sperm development also differs in different polychaetes. In the nereids, removal of the brain causes the precocious appearance of sperm in the coelom, which can be prevented by subsequent brain implantation. In *Arenicola*, hormones secreted by the brain control spermatogenesis in two ways.[151] Removal of the brain from almost mature worms prevents the separation of the spermatozoa from the sperm morulae, and as in the female there is a hormonal stimulation of maturation. In addition a hormone secreted by the brain may stimulate the proliferation of spermatocytes from the gonad, for if an adult worm is stripped of spermatozoa, the testis responds with an outburst of mitotic activity, which is prevented if the brain is removed. Mitotic activity in decerebrate worms can be induced by the injection of extracts of brains taken from worms that have been stripped of their gametes. The secretory activity of the brain is thus controlled in part by the quantity of coelomic spermatocytes, an unequivocal example of a feedback mechanism.[151]

The control of epitoky

Removal of the brain from a young *Nereis* is followed by premature epitokal metamorphosis, as well as by precocious and accelerated oocyte growth. Implantation of a brain from a young individual into a decerebrate mature *Nereis* not only prevents epitoky but also prolongs the life of the worm. The brain of a young *Nereis* thus produces a hormone which inhibits epitoky.

However, as with the control of gametogenesis, the brain hormone is not purely inhibitory, but plays an active part in promoting epitoky once the initial inhibition has been lifted. Thus if the brain of *Platynereis dumerilii* is removed from an individual whose oocytes are larger than 100 μ in diameter, epitokal metamorphosis is completed in 5–7 days, rather than the normal time of 8–15 days. But if the oocytes are less than 80 μ in diameter at the time of brain removal, then epitokal metamorphosis is abnormal and the animal dies within 12 days. Brains taken from very young females with no visible oocytes and implanted into headless individuals with oocytes 100–120 μ in diameter will retard further oocyte development in the hosts, prevent further epitokal metamorphosis, but induce segment regeneration. If these implanted brains are then *removed* after two weeks, metamorphosis continues to completion. A brain from a small, very young *Platynereis* inhibits metamorphosis completely when implanted into the parapodium of an older individual, but when implanted into a decerebrate fragment of 30 segments, metamorphosis is initiated and proceeds normally. Thus, exactly like the hormone from the brain which controls oocyte growth and vitellogenesis, the brain hormone controlling epitokal metamorphosis is inhibitory in high concentration but stimulatory in lower titre. Normal metamorphosis must therefore depend upon the gradual decrease in hormone production by the brain.[125]

The relationship between gametogenesis and epitoky

The endocrine control of gametogenesis and epitokal metamorphosis is so similar that it seems very likely that only one hormone from the brain controls both processes. Preliminary purification of nereid brain extracts shows that the water-soluble fraction inhibits both gametogenesis and epitoky,[81] but whether this is because only one hormone is involved, or that at least two hormones with similar chemical properties separate together in the purification process, must await further and more detailed analysis.

That brain removal from a young *Platynereis* causes accelerated but abnormal gametogenesis and epitoky, whereas brain removal from older individuals allows both to proceed normally, is often taken to mean that the two processes have similar threshold responses, and must therefore be reacting to a single hormone. But the possibility cannot be eliminated that two hormones are involved, closely synchronized both in production and action.

In *Perinereis cultrifera*, homogenates of the coelomic contents of females with oocytes 200–250 μ in diameter injected into females with oocytes 30–60 μ in diameter sometimes accelerate metamorphosis (in 6 individuals out of 27 in one experiment) without any concomitant effect upon oocyte growth.[80] This could mean that in the injected worms, the production of gametogenic hormone is unaffected by the homogenate, whereas the production of metamorphic hormone is prevented, so allowing metamorphosis to proceed. Such a differential effect would imply the existence of two distinct hormones. But in similar experiments with *Platynereis dumerilii*, none of the injected worms are affected in any way by the homogenate.[126] In both nereid species, when gametogenesis is prevented by x-irradiation or starvation, epitokal metamorphosis occurs more or less normally.

At present, the most economic hypothesis is that a single brain hormone controls both gametogenesis and epitoky in nereids. The two processes must be intimately linked; but in insects the equally important association of vitellogenesis and accessory gland development, both conclusively shown to be controlled by the same hormone (p. 136), provides a precedent for a single hormone affecting two separate but related processes.

Control of stolonization

In syllids, the brain is quite unequivocally *not* the source of hormones controlling stolonization. Instead, removal of the proventriculus induces premature stolonization, which suggests that this organ (or some structure associated with it) may produce an inhibitory hormone in young animals. But reimplantation of the proventriculus does not inhibit the precocious stolonization induced by its removal, so a hormonal function for the organ must at present be open to doubt.

Reproduction, growth and regeneration

In most polychaetes, somatic growth is at first rapid, with a high rate of segment proliferation continuing until about half the usual segment number is achieved. Then both segment proliferation and enlargement slow down considerably, and eventually stop altogether. Vitellogenesis and the final maturation of both oocytes and spermatocytes occur during the latter period of somatic growth. If segments are amputated during this time regeneration will not occur. It might therefore be supposed that at an appropriate time, nutrients are switched from somatic growth to gametogenesis and that this switch is manipulated by hormones. The epitokal transformation must also use considerable energy. It is possible that all polychaetes at one time metamorphosed in a manner similar to the nereids, but have since become neotenous, that is, reproduction is associated with a juvenile body form, with a consequent saving in energy. On this interpretation the nereids must have retained the original transformation because the advantages of a different reproductive habitat outweighed in this group the disadvantages of epitoky.

The brain of a mature nereid will not induce regeneration when implanted into a decerebrate young individual. Thus production of growth and regeneration hormone by the brain must cease as the individual mature. But a brain from a young nereid will induce segment prolifera tion in decerebrate mature worms, and regeneration in amputated mature worms. So the tissues of mature nereids retain the potentiality for growth and regeneration, but the potential is not realized because of the normal absence of the appropriate hormone. Moreover, the implanted young brain also inhibits gametogenesis and epitoky in its mature host; both processes go on normally if the implanted brain is subsequently removed (p. 59). Is the growth and regeneration hormone therefore identical with the gametogenesis and metamorphosis hormone(s)?

In *Nereis diversicolor*, the secretion by the brain of growth-promoting hormone and of oocyte-inhibiting hormone ceases at about the same time, suggesting that only one hormone is involved in both processes. But when segments are amputated early in the second, decreasing growth phase, oocyte development is not inhibited although segment regeneration occurs. It is suggested that this result indicates the existence of distinct growth-promoting and oocyte-inhibiting hormones: the oocytes develop in a normally decreasing inhibitory hormone concentration, although the growth hormone titre is increased to facilitate segment regeneration. But in *Platynereis dumerilii*, segment regeneration *does* retard oocyte development, a result consistent with a single hormone hypothesis. It is possible, of course, that in *Nereis diversicolor* the oocytes are inhibited by a growth hormone concentration higher than that achieved during segment regeneration in older individuals, such a concentration being present in young actively growing worms. So the difference between *Nereis* and *Platynereis could* be due to variety in the oocyte threshold of response to a single hormone, rather than to the existence of different hormonal mechanisms in the two species. But the problem will not be definitely resolved without further experiments and the use of much more highly purified extracts of brain hormones.

GROWTH AND REPRODUCTION IN OLIGOCHAETES

Regeneration in oligochaetes, like that in polychaetes, seems to be controlled by neurosecretory hormones produced by the brain, although much of the evidence depends more upon histological changes in the cerebral neurosecretory cells than upon precise experimentation.[129]

In *Eophila dollfusi*, for example, removal of the brain followed by amputation of caudal segments, causes the worm to enter diapause during which regeneration occurs. Amputated worms with intact brains heal the wound but do not regenerate. This might suggest that the brain produces a hormone which *inhibits* regeneration. But during normal diapause, material accumulates within the neurosecretory cells of the brain and the ventral ganglia. When segments are amputated, the accumulated neurosecretion is rapidly released and segment regeneration occurs.[102]

These seemingly contradictory results could be explained by assuming that removal of the brain from a non-diapausing individual does not have an endocrine effect upon regeneration, but rather induces a period of diapause during which reproductive activity ceases, the reproductive organs regress, and a state favourable for regeneration is brought about; regeneration itself being stimulated by neurosecretory hormones from the ventral ganglia. It has even been suggested that sexual reproduction and regeneration might depend upon the activity of two distinct types of neurosecretory cell in the ventral ganglia, which cannot function at the same time. The activity of one cell type suppresses the activity of the other, overall control being exercised by the brain. This hypothesis could explain the antagonism between regeneration and sexual reproduction, but histology must be supported by decisive experimentation before valid conclusions can be established.

In the earthworm *Eisenia foetida*, removal of the brain arrests immediately gonadal development, and results in the disappearance of structures, such as the clitellum, associated with copulation and egg laying. The brain contains several types of neurosecretory cells, and histological cycles in the cells can be correlated with normal reproductive development. Removal of the sub-oesophageal ganglion will also prevent egg-laying, but only after an interval of time, and this operation probably interferes with the passage of cerebral neurosecretion from the brain. Re-implantation of the brain into a decerebrate worm stimulates the production of spermatids from spermatogonia.[127, 128]

Although endocrine mechanisms in the oligochaetes have been far less intensively studied than those in the polychaetes, it can at least be concluded that a major difference exists between the two groups: in the polychaetes, both growth and reproduction are initially *inhibited* by neurosecretory hormones; in the oligochaetes, neurosecretions stimulate the two processes.

REPRODUCTION IN LEECHES

The Hirudinea are the third major group of annelids. Leeches, like earthworms, are monoecious (hermaphrodite), and, although gametogenesis occurs simultaneously in both ovaries and testes, oogenesis lags slightly behind spermatogenesis.

The leech testis shows a cycle of activity correlated with the seasons. Spermatocytes are budded as single cells from the thin walls of the testis and are released into the fluid filled testicular lumen. Here mitotic and meiotic divisions produce clusters of spermatozoa attached by their heads to a central cytoplasmic syncytium. In *Hirudo medicinalis* many mature sperm clusters are present during August, and are completely absent from November to April.[120]

It was first suggested that this testicular development was hormonally controlled when it was found that certain neurosecretory cells (the α-cells) in the brain decreased both in number and in their content of neurosecretion during the latter part of the year, reaching a minimum in December, but then increased to a maximum from April to June.[120]

The possibility of such a causal relationship between the cyclic production of neurosecretion and testicular activity needs to be verified by experiments involving the removal and reimplantation of the brain. The brain of the leech is very difficult to remove through the outer body wall, but fortunately the animals can be turned inside-out, and the brain removed through the gut wall. Naturally, the effects of the operation are compared with those in control animals, also turned inside-out, but from which the brain is not removed.[120]

In such decerebrate animals, the total number of gamete clusters becomes very much less than in the controls. Homogenates of brains taken from leeches in early February injected into decerebrate individuals markedly increase the total number of gamete clusters.[120] It can be concluded, therefore, that the leech brain produces a gonadotrophic hormone which has a positive effect upon spermatogenesis (Fig. 4.12). This situa-

Treatment	Average numbers of gametes (per field of view)	
	First half of development	Second half of development
Untreated	306	393
Brain removed	307	69
Brain removed *plus* injection of brain homogenate	296	237

Fig. 4.12

tion is therefore similar to that in the oligochaetes, and differs from that in the polychaetes, a conclusion which is in line with the supposedly close relationship between oligochaetes and leeches.

REPRODUCTION AND ENVIRONMENT

The spawning of polychaetes is often related to environmental events, and there is usually some degree of co-ordination in spawning between the members of a population. In certain species, this has acquired the strictest periodicity. For example, in the palolo worm, *Leodice viridis*, which lives in the reefs of the Southern Pacific, as the day of the last lunar quarter of the October–November moon dawns, the posterior half of the body breaks off and swims to the surface to spawn. The anterior end of the worm regenerates its missing portion and spawns again the following year. The times of swarming are so predictable that they are included in the local Fijian calendar. Of course, in many other so-called palolo worms, the periodicity is nothing like so precise as in *Leodice*.

In the vast majority of polychaetes, spawning is not so well co-ordinated as in *Leodice* and other species. But even so, some degree of co-ordination is necessary. In *Platynereis dumerilii*, the breeding season extends from March to October in the Mediterranean and there is a monthly periodicity of surface swarming, the largest numbers appearing at the time of the new moon and the smallest at the full moon. *Platynereis* swarms on the night following the completion of metamorphosis, so the time at which metamorphosis *begins* will determine the time of swarming. Since metamorphosis takes about 1 to 2 weeks to complete, it must be initiated at about the time of a full moon.[124]

It might be expected then that increasing photoperiod could be the trigger for the initiation of metamorphosis. When *Platynereis* is kept in constant light, the periodicity of swarming is eventually abolished. Exposed to 12 hours light and 12 hours dark in every 24 hours, followed by 6 days constant light at the end of every month, *Platynereis* will spawn 16–20 days after the end of the period of constant light. It appears, then, that increased photoperiod initiates metamorphosis, presumably by causing the cerebral neurosecretory cells to stop producing hormone (p. 58).[124]

But *Platynereis* will spawn even though the previous critical full moon was obscured by cloud. This together with the maintenance of swarming periodicity in constant illumination for a period of up to 3 months, suggests that an endogenous cycle becomes imprinted in the worms, perhaps by exposure to exogenous lunar cycles in early life.

Blinded worms can be synchronized to an artificial photoperiod in precisely the same way as normal worms. The eyes therefore are not the only pathway which transmit information about the environmental photoperiod to the cerebral neurosecretory cells. Exactly the same phenomenon is seen in insects and vertebrates (Chapter 14). It is possible that the neurosecretory cells themselves, or perhaps some nervous centre in close proximity, react directly to environmental photoperiod.[124]

Another factor in the co-ordination of swarming may be the production of pheromones—released in a swarming individual to cause other mature individuals to swarm at that precise time. But pheromone action is much better known in insects, and its discussion will not be pursued here (see p. 231 *ff*).

The influence of a single neurosecretory hormone upon growth and regeneration in many polychaetes, its withdrawal allowing the alternative of reproduction, recalls the situation in *Hydra* (p. 35). But the results of those experiments which underline the importance of the target in its reaction to the same hormone concentration, form a notable addition to our understanding of endocrine mechanisms in all animals. Although the sites of production of the hormone(s) are not known precisely, a very firm foundation now exists for future research. Moreover, although not examined in as much detail, the endocrine mechanisms in oligochaetes and leeches are clearly different from those in polychaetes and suggests the diversity of hormonal control systems which possibly exist within the annelids as a whole. Such diversity is a feature of endocrine mechanisms in other groups, for example the arthropods, where they are discussed in more detail. But first, in the next chapter, the evidence for endocrine mechanisms in the molluscs will be assessed.

5

Endocrine Mechanisms in the Mollusca

As has been shown in Chapter 2, neurosecretory cells have been identified histologically within the nervous systems of coelenterates, nematodes, annelids, etc., and subsequent experiments have suggested that these cells are the source of hormones controlling development. Because of the nature of the neurosecretory systems in these animals, it cannot be proved unequivocally that such cells actually produce the hormones. But the combination of histology and experiment provides good circumstantial evidence for such a view.

In the Mollusca, the problem is more acute. Many neurones within the ganglia of the nervous system stain with dyes, such as paraldehyde-fuchsin and chrome haematoxylin-phloxine, which have been used to characterize neurosecretory cells in other animal groups.[246, 100] But since these staining reagents are not specific for neurosecretory material, histological criteria alone provide insufficient evidence that the neurones are neurosecretory. The stained inclusions within such neurones sometimes show cyclic changes associated with other events in the animal, but this does not provide additional evidence for a neurosecretory function, since cyclic changes in neuronal inclusions, such as lysosomes or glycogen granules, are not improbable. Even if some of these neurones are neurosecretory cells, it is impossible to decide whether the cyclic changes in their histology indicate a neurosecretory control over some developmental process in the animal, or whether the neurosecretory cells *and* the developmental process are both reacting to some other factor. The neurosecretory cells could even be reacting to the process itself in order to control some quite different event within the animal.

The way to resolve this difficulty, of course, is to relate such histological observations to experiments in which the neurones suspected of being neurosecretory are removed and subsequently reimplanted and the effects of the operations noted. But in fact, few attempts have been made to determine experimentally the hormonal functions of suspected neurosecretory cells in the molluscs. To a large extent, this lack of experimentation is due to the inherent difficulties of removing small numbers of cells buried deeply within nerve ganglia. Total removal of any ganglion often results in gross malfunction of the operated individual followed by death. In only a few instances have histological studies been accompanied by surgical removal of ganglia. This account of molluscan endocrinology will be confined to these examples, since they demonstrate that experimentation rarely produces results consistent with hypotheses established by previous histological observations on supposed neurosecretory cells. Such contradictions should be borne in mind if reference is ever made to the wealth of literature on the histology of molluscan neurosecretion.

NEUROSECRETION IN GASTROPODS

In the pond snail, *Lymnaea stagnalis*, the cerebral ganglia each contain three groups of neurosecretory cells (Fig. 2.7),[190] the histology of which changes with the seasons. The cells of the medio-dorsal and latero-dorsal groups stain in a similar manner, and contain very little neurosecretory material in winter.[158] But in spring, neurosecretion accumulates in both the axons and the perikarya of the cells, the rate of accumulation slowing down during summer and autumn (Fig. 5.1). These changes are said to indicate maximum production, transport and release of neurosecretion during April and May by the cells of the medio-dorsal and latero-dorsal groups. But such an interpretation should be viewed with caution, since it is difficult to relate static histological pictures to an essentially dynamic process (see p. 139 and Fig. 7.30).

The caudo-dorsal groups of neurosecretory cells in the cerebral ganglia (Fig. 2.7) stain differently from those in the other groups, which might indicate the production by these cells of a different neurosecretory hormone. The cells go through a seasonal cycle of histological change similar to that in the medio-dorsal and latero-dorsal cells (Fig. 5.1), and they are also considered, in consequence, to be inactive in winter, and highly active in spring.[158]

Lymnaea is hermaphrodite, possessing an ovotestis. The male component of the ovotestis shows a clear periodicity, with a peak of spermatogenesis during April and May. But the proportions of early oocytes, fully grown oocytes surrounded by follicle cells, and degenerating oocytes remains constant throughout the year. Clearly, as egg masses are laid, there is continual development and replacement of new oocytes. However, periodicity in spermatogenesis and its absence in oogenesis suggests that the development of the two components of the ovotestis is controlled by different mechanisms. What may these mechanisms be?

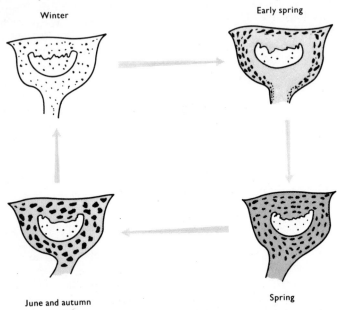

Fig. 5.1 Histological changes in the medio-dorsal and latero-dorsal neurosecretory cells of the pond snail, *Lymnaea stagnalis*. During winter the cells contain only a small amount of neurosecretory material, consisting of small spherical granules dispersed throughout the cytoplasm. In early spring, the quantity of neurosecretory material begins to increase: flakes of material occur near the peripheries of the cells with very small granules near the nuclei. In late spring, the flakes of material increase in number and fill each cell. During summer and autumn the flakes seem to become more compact, occupying less of the cytoplasm of each cell. These histological changes may represent a cycle of synthesis and release of neurosecretory hormone during early spring (but see Fig. 7.30). (After Joosse[158])

If the histology of the medio-dorsal and the latero-dorsal neurosecretory cell groups in the cerebral ganglia has been interpreted correctly, these cells are maximally active during the peak of spermatogenesis. In young animals, the caudo-dorsal neurosecretory cells do not show this presumed activity, and no egg masses are produced in such animals.

Consequently, it is possible that the medio-dorsal and the latero-dorsal neurosecretory cells control spermatogenesis, and the caudo-dorsal cells (with different staining properties) control oviposition and perhaps oogenesis also. How does this possibility stand the test of direct experimentation?

When the cerebral ganglia of *Lymnaea* are completely removed, feeding activity, lung ventilation, copulation and egg-laying all stop, and the snail moves about slowly and only for short distances. The operation clearly has very drastic effects upon the life of the animal. Joosse has been able, by the most delicate manipulative techniques, to remove the neurosecretory cell groups from the brain without obvious damage to the latter.[158] But even so the destruction of even a few neurones in the vicinity of the neurosecretory cells could affect the general viability of the snails. However, removal of any or all of the neurosecretory cell groups from the cerebral ganglia of *Lymnaea* has little effect upon the

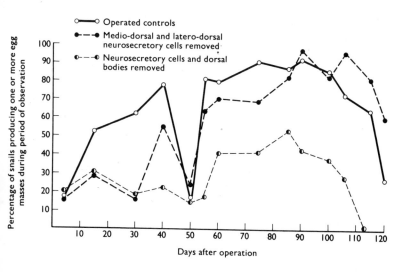

Fig. 5.2 Effects of removal of the medio-dorsal and the latero-dorsal neurosecretory cells and the dorsal bodies on egg mass production in the pond snail, *Lymnaea stagnalis*. Removal of the neurosecretory cells results in a lowered production of egg masses during the first half of the experimental period, but during the second half production of egg masses is not significantly different from that in the control animals. When the neurosecretory cells *and* the dorsal bodies are removed, the rate of production of egg masses remains low throughout the experimental period. These results suggest that the dorsal bodies, rather than the neurosecretory cells, control the production of egg masses in *Lymnaea*. (After Joosse[158])

development of the ovotestis (Fig. 5.2). Thus in spite of so-called activity cycles in these neurosecretory cells, closely correlated with gameto-genesis, there is no *direct* relationship between the two.

But many gastropods have other endocrine organs, apparently developed from glial cells (which are a kind of supporting cell found in all animals with a complicated central nervous system) and the peri-neurium of the cerebral ganglia, which are often connected with the ganglia (Fig. 2.7). These organs are the dorsal bodies, usually consisting of paired medio-dorsal and latero-dorsal bodies.[158]

In *Lymnaea*, the cells of the dorsal bodies have small nuclei and very little cytoplasm from October to January. But then the nuclei enlarge and the quantity of surrounding cytoplasm increases. Thus the cells of the dorsal bodies undergo changes suggesting increased activity just before the ovotestis is stimulated.

When the medio-dorsal bodies are removed, the number of egg masses produced by the operated individuals is considerably reduced. When both medio-dorsal and latero-dorsal bodies are removed, both the number and the size of the egg masses are reduced (Fig. 5.2). It is suggested, therefore, that the dorsal bodies produce a hormone, or hormones, which control both ovulation and oogenesis in *Lymnaea*.[158] Unfortunately, there has yet been no subsequent implantation of the dorsal bodies or even the injection of extracts, so this postulated endo-crine control must remain tentative at present. Removal of the dorsal bodies has no obvious effect upon spermatogenesis, and the question of a differential endocrine control over the ovotestis must also remain open.

When the dorsal bodies are removed, the neurosecretory cells in the cerebral ganglia are invariably damaged, so a functional relationship between these cells and the dorsal bodies is difficult to establish. Recent electron microscope studies of the system show no neurosecretory axons from the cerebral ganglia in the dorsal bodies. So if cycles in the amount of inclusions in the neurosecretory cells are of any significance in con-trolling the activity of the dorsal bodies, the effect must be indirect, presumably through the blood.

In the snail, *Helix aspersa*, and the slug, *Arion*, neurosecretory cells have been identified in the optic tentacles, particularly surrounding the tentacular ganglia.[180] When the optic tentacles of *Arion* are removed, egg production is increased, to be reduced again to normal after the subsequent injection of tentacle extracts. The injection of extracts of cerebral ganglia into *intact* animals also increases egg production, whereas tentacle extracts have no effect.[220] These results suggest that the tenta-cular neurosecretory cells normally inhibit egg production, while those in the cerebral ganglia stimulate the process. Such a dual hormone system could exercise an exact control over egg production, and it might be supposed that the neurosecretory cells in the optic tentacles would react to environmental stimuli, correlating reproduction with the appropriate conditions. In *Helix*, removal of the tentacles is followed by changes in the hepatopancreas which might suggest that the tentacular neurosecretory cells in this species regulate metabolism. There is, however, still some dispute about the exact nature of these presumed neurosecretory cells in the tentacles of *Arion* and *Helix*[246, 181]: some at least could be mucus-secreting cells. Moreover, the endocrine control of egg production in these stylommatophoran pulmonates appears to be quite different from that of a basommatophoran like *Lymnaea*. Much more detailed experimental work on the endocrine control of reproduction in many more pulmonates is necessary before any general principles can be established.

Much more convincing evidence exists for the control of water balance in *Lymnaea* by neurosecretory hormones.[191] When the snails are kept in hypertonic salt solution, material disappears from the neurosecretory cells of the lateral lobes of the cerebral ganglia. Neurosecretion accumulates in these same cells when the snails are kept in distilled water. These observations suggest that when the snail is in danger of losing water due to osmotic forces, a hormone is released from the cerebral neurosecretory cells to conserve body water; that is, an antidiuretic hormone. But again these histological changes should be interpreted with caution. Severe osmotic stress can induce changes in many cells and tissues, and the neurosecretory cells may be reacting passively to hypertonicity, and not themselves attempting to regulate it.

In fact, in *Lymnaea*, the removal of the cerebral ganglia has little effect on the control of body water.[191] Of all the ganglia, only when the pleural ganglia are removed is there any remarkable change in the snail's water balance: the body swells up owing to an accumulation of water. Removal of the left pleural ganglion alone has little effect; removal of the right pleural ganglion alone results in slightly less swelling than does bilateral extirpation. Reimplantation of the right pleural ganglion prevents the massive accumulation of water that follows its removal, and the single injection of an homogenate of right pleural ganglion into a normal snail causes a rapid reduction in weight.[191] The results of all these experiments lead to the firm conclusion that the right pleural ganglion in *Lymnaea* produces a diuretic hormone.

Both pleural ganglia in *Lymnaea* contain neurosecretory cells, in about equal numbers. There is thus no *histological* basis for the difference in diuretic activity between the two ganglia. But in the same way that the different histology of neurosecretory cells may not reflect different activities, cells with the *same* histology may actually produce and release hormones at very different rates (see Fig. 7.30).

NEUROSECRETION IN LAMELLIBRANCHS

In the edible mussel, *Mytilus edulis*, material accumulates in the neuro-secretory cells of the cerebro-pleural and visceral ganglia during the period of gamete maturation. Discharge of the gametes is preceded by the disappearance of neurosecretion from the cells.[194] Removal of the cerebro-pleural ganglion has no effect upon gamete maturation but apparently accelerates gamete discharge. Removal of the visceral ganglion retards oviposition.[195]

If it is assumed that ganglion removal has no traumatic effect upon *Mytilus*, the relationship between neurosecretory cell histology and the results of extirpation are equivocal, to say the least. If the sudden dis-appearance of neurosecretion means the release of a hormone which causes gamete discharge, then the acceleratory effect of removal of the cerebro-pleural ganglion is inexplicable. It is likely that experiments involving reimplantation of the nerve ganglia, or the injection of homo-genates, would produce results which would clarify the situation enormously.

In the zebra mussel, *Dreissena polymorpha*, there is an even closer parallel between the histology of neurosecretory cells in the cerebro-pleural and visceral ganglia and the reproductive cycle. In *Dreissena*, both neurosecretory and reproductive cycles begin at the end of summer. The histology of the neurosecretory cells suggests a slight discharge of material at the end of autumn or beginning of winter, and simultaneously there is a spurt in oocyte growth. In the spring and summer, maximum discharge of neurosecretion is correlated with intensive oocyte growth and spawning.[3] Neurosecretion is said to be discharged when the amount within the cells diminishes.

But when the cerebro-pleural ganglion is removed in March—at the onset of oocyte growth—the oocytes still grow to maturity. Spawning can also take place some weeks after removal of the cerebro-pleural ganglion. These experimental results contradict completely the relation-ship between neurosecretion and oocyte development and spawning, established from interpretations of neurosecretory cell histology. Re-moval of the visceral ganglion results in rapid death.[3]

In both gastropod and lamellibranch molluscs, therefore, well estab-lished correlations between cycles of activity in presumed neurosecretory cells and reproductive or physiological events, do not survive the rigours of experimental method. On the contrary, more real information about endocrine mechanisms in these groups has been obtained from the few methodical experimental studies which have been carried out, than is contained in the large descriptive literature on supposed neurosecretory phenomena. In any animal, the interpretation of cycles of activity in neurosecretory cells is very difficult even when experiments have indicated the controlling effects of hormones at different times. Attempts to diagnose neurosecretory activity histologically without correlated experiments can lead to the most erroneous conclusions.

THE CONTROL OF SEXUAL MATURITY IN CEPHALOPODS

The cephalopod molluscs include the squids and octopi, which are perhaps the most highly developed of all invertebrate animals. They have a large and complicated brain (Fig. 2.10) and their eyes are functionally equivalent to those of the vertebrates. Both octopi and squids are remarkably intelligent creatures, and are very popular animals for use in experiments upon learning and behaviour. In fact, the best account of the endocrine control of sexual maturation in *Octopus* arose as a secondary consequence of such experiments.

In the immature female *Octopus*, the optic glands (Fig. 2.10) are small and pale, but increase in size some ten times and become bright orange as the ovaries enlarge.[268] M. J. Wells, investigating the effects upon learning of the removal of different parts of the brain of *Octopus*, or of cutting connections between its various parts, discovered that a certain proportion of his experimental females became precociously mature, and these all had enlarged optic glands. Further investigations showed that three kinds of operation upon the brain causes enlargement of the optic glands: (a) removal of the subpedunculate lobes (Fig. 5.3); (b) cutting the nerves between these lobes and the optic glands; and (c) blinding the animals by cutting the optic nerves, or by cutting the optic stalks distal to the optic gland and removing the optic lobes (Fig. 5.3). Ovarian enlargement is more rapid after operations (a) and (b) than after (c). Lesions elsewhere in the brain have no effect upon maturation, and do not increase the size of the optic glands: these can be considered control experiments for the three operations which do cause precocious maturation. Moreover, when the optic glands are removed, no matter what other lesions are made in the brain, precocious maturation does not occur. Finally, when the lesions that cause enlargement of the optic gland are made unilaterally, the gland on the other side is unaffected.[268]

What conclusions can be drawn from these results? First, it seems clear that a hormone from the optic glands controls ovarian enlargement: this is caused largely by yolk deposition in the existing oocytes, and so the hormone is vitellogenic and presumably acts upon the follicle cells around the oocytes. Although optic glands have not been reimplanted into animals from which the organs have been removed, their virtual isolation by nerve sections can be considered equivalent. Secondly, the

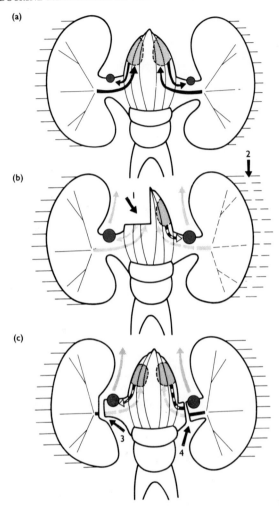

Fig. 5.3 Hormonal control of sexual maturity in *Octopus*. (a) In an immature, unoperated animal, the activity of the optic glands is suppressed by inhibitory nerves. It is likely that changes in photoperiod govern the inhibitory centres in the brain and hence the secretion of the optic glands. (b) Operations (shown by black arrows) which cause enlargement—and secretion—of the optic glands and result in development of the gonads: (1) removal of the source of the inhibitory nerve supply; (2) section of the optic nerves, causing blindness. (c) Operations (shown by black arrows) which cause enlargement of the optic glands and gonads, and which eliminate the possibility of stimulatory innervation of the glands; (3) removal of the optic lobes; (4) section of the optic tract. Compare Fig. 2.10. (After Wells and Wells[268])

effects of unilateral nerve sections suggests that optic gland enlargement is not the result of any intermediary endocrine control: the contra-lateral optic gland would also enlarge if this were so. Consequently, the optic glands must be inhibited in the immature female by a nervous centre in the subpedunculate lobes of the brain; these centres are themselves controlled by nervous impulses coming from the eyes. This hypothesis would explain the more rapid ovarian enlargement that occurs after removal of the subpedunculate lobes or section of the optic tract proximal to the optic glands, when compared with the effects of section of the optic nerves or optic tract *distal* to the glands (see Figs. 2.10 and 5.3).

A similar control over testis development by the optic glands in the male *Octopus* can be demonstrated, except that the testes apparently react to a lower titre of hormone. Furthermore, the males normally mature much earlier than the females, and when the optic glands are removed, the testes can actually regress. This suggests that the development of the testes needs a continuous supply of optic gland hormone; it is likely that the same applies to ovarian development.

The secretory activity of the optic glands is thus inhibited in the immature octopus by nervous means. There is no evidence for the intervention of neurosecretory hormones in this inhibition. Nor is it clear how the nervous inhibition is lifted during normal maturation. Wells suggests that special mechanisms for delaying maturation are found in the arthropods and vertebrates as well as in the cephalopods, and that such mechanisms are important only in these highly developed animals because the central nervous system must develop its full potentialities before the gonads become mature. On this view, the control of endocrine activities by the central nervous system is not surprising. But the cephalopods differ from the arthropods and vertebrates in their *direct* nervous control over the endocrine organs responsible for maturation. It may be that the development of the full potentiality of the central nervous system is only one aspect of the general antagonism between somatic and reproductive processes found in many animals already discussed in Chapter 4.

THE BRANCHIAL GLANDS OF CEPHALOPODS

In the octopus, these large parenchymatous glands, closely associated with the gills (p. 19), are impossible to remove without causing the death of the animal within about 2 days. The operated individuals lose blood from the accessory hearts (Fig. 2.11). But when the glands are destroyed by cautery, the octopus takes very much longer to die, and in the meantime shows loss of appetite, lack of growth, muscular inertia, decreased nervous activity and many other symptoms of a generally reduced vitality. Unfortunately, branchial gland implants do not survive, and replacement therapy for animals without their glands has to take the form of repeated injections of branchial gland extracts. But although this treatment alleviates to a certain extent many of the symptoms of branchial gland removal, the operated animals still do not live for very much longer than do the uninjected individuals without branchial glands.[258]

The branchial glands are well vascularized and have no nervous supply. Their cells produce secretions which apparently pass into the blood (Fig. 2.11). Histologically, therefore, there are good grounds for regarding these glands as endocrine. Their secretion is soluble in alcohol and survives boiling.[258]

If the branchial glands are considered to be endocrine on the more uncertain basis of the experimental results, then their functions would be analogous to those of adrenocortical tissue in the vertebrates. The branchial glands of cephalopods would also have the distinction of being the largest endocrine glands in the invertebrates, and the only mesodermal ones other than those in the arthropods. But until pure preparations of their secretion(s) can, on injection, offset completely the effects of destruction of the branchial glands, their endocrine nature must be considered doubtful.

6

Endocrine Mechanisms in the Insecta—I

The invertebrate animals, the endocrinology of which has already been discussed, all possess dispersed neurosecretory cells, are without neuro-haemal organs (or else the glands are very poorly developed), and with the exception of the cephalopod molluscs are devoid of epithelial endo-crine glands. Consequently, only rarely can the sources of hormones be located precisely.

But in the insects, some neurosecretory cells are grouped conspicu-ously within the brain and are associated with particularly well developed neurohaemal organs. At least two pairs of epithelial endocrine glands are present (Chapter 2). Insects are therefore much more favourable animals for the investigation of endocrine mechanisms than any of the less highly organized invertebrates. The great economic importance of insects, together with the relative ease with which many species can be reared in the laboratory, are added inducements for such investigations. As a result, more is known about the endocrinology of insects than any other invertebrate group.

Because of the nature of the insect endocrine system, experiments can be more sophisticated and produce more definite conclusions than any described previously. An insight into the mechanisms of hormone action has been gained, and some insect hormones have been isolated, purified and identified chemically (Chapter 12). A full discussion of insect endocrinology would occupy more space than this book can provide: the accounts which follow have been considerably condensed.

GROWTH, MOULTING AND DIFFERENTIATION

In insects, growth and differentiation are profoundly influenced by the cuticle which must be detached before the epidermal cells can grow and divide, and shed before the individual can increase in size. This moulting of the cuticle usually occurs a fixed number of times in any species. The interval between moults is called a stadium, and the form that the insect takes in any stadium is called an instar.

The overall change in form between the first larval instar and the adult is called metamorphosis. This may be slight, as in the primitive wingless, or apterygote, insects like the silverfish, involving merely the appearance of scales, coxal styles and external genitalia (Fig. 6.1). It may

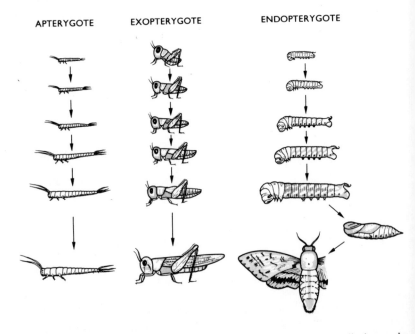

Fig. 6.1   Metamorphosis in insects. Metamorphosis is the overall change in form between the first larval instar and the adult. Change in form only becomes apparent during the periodic moults. In the primitive apterygote insects metamorphosis is slight, consisting in the progressive appearance of scales, coxal styles and external genitalia. The exopterygote insects show greater changes: wing pads appear in the later larval instars, developing into functional wings at the last moult; the external genitalia also develop progressively. In the endopterygote insects, the larval instars are similar and metamorphosis takes the form of a sudden transformation of the last larval instar to pupa and hence to adult.

be more pronounced as in the exopterygote insects (cockroaches, grass-hoppers, bugs etc.), where the appearance of wing pads in the later larval instars and their transformation into functional wings at the last moult may be more obvious signs of progressive differentiation than the development of genitalia (Fig. 6.1). Finally, metamorphosis may take the form of a sudden transformation of the larval stage into a form—the pupa—quite unlike it in appearance with the subsequent emergence of the adult (Fig. 6.1). This complete or indirect metamorphosis occurs in the endopterygote insects—beetles, ants and bees, butterflies and moths, flies etc.

In endopterygote larvae, bodily form, mouthparts and sense organs (particularly in the absence of compound eyes) are all different from those in the adult, indicating the quite separate ways of life of the larva and adult. The pupal stage is that in which many of the purely larval struc-tures are destroyed and replaced by those of the adult. But the precursors of the adult structures are present in the larvae in the form of imaginal buds, nests of cells separate from the functioning larval tissues. In some endopterygotes, wings, legs and mouthparts are formed from these

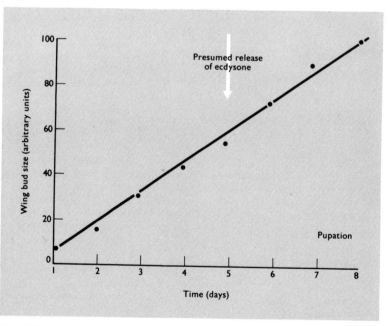

Fig. 6.2 Increase in size in the wing buds of the blowfly, *Calliphora*, during the last larval instar. The rate of growth of the wing buds is constant whether or not ecdysone is present.

imaginal buds which undergo some development even in the larval stages (Fig. 6.2). But in other endopterygotes, all the adult body, including the epidermis, is developed from imaginal buds and cells. In the exopterygotes and the apterygotes, on the other hand, the daughter cells which give rise to larval tissues at an early moult are the direct precursors of those which eventually form the adult tissues (Fig. 6.3).

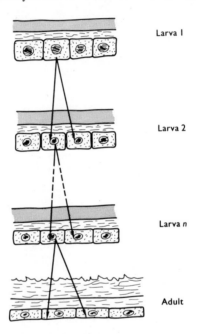

Larva I

Larva 2

Larva *n*

Adult

Fig. 6.3 Epithelial cell lineage in exopterygote insects. Daughter cells of the epithelium of the first instar larva compose the epithelium of the second instar, and so through successive instars to the adult. Compare Fig. 6.22.

A comprehensive theory of the endocrine control of moulting, growth and differentiation in insects must explain these different metamorphoses.

Endocrine control of development

Current views on the endocrine control of insect development can be summarized as follows. A hormone produced in the neurosecretory cells of the brain is carried by the neurosecretory cell axons to the corpora cardiaca, from where it is released into the blood. This hormone then stimulates the thoracic glands (or their equivalent) to produce a hormone which causes the epithelial cell to begin the processes which lead

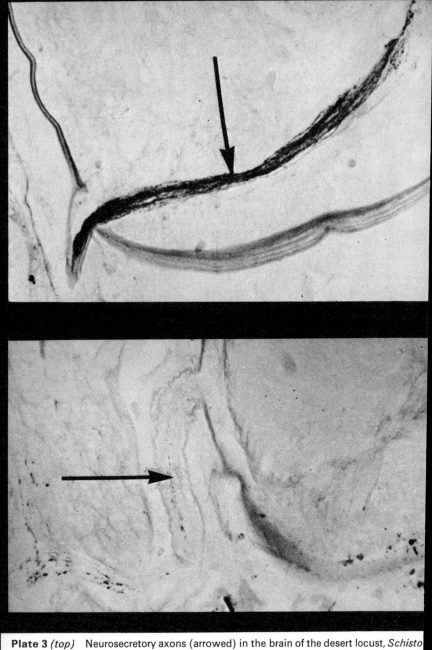

Plate 3 *(top)* Neurosecretory axons (arrowed) in the brain of the desert locust, *Schisto gregaria*, after 30 minutes low frequency stimulation (10 volt, 5 msec pulses at 40/mi the central nervous system. The axons contain a relatively large amount of neurosecreti

Plate 4 *(bottom)* Neurosecretory axons (arrowed) of the desert locust after 30 mi high frequency stimulation (10 volt, 5 msec pulses at 40/sec) of the central nervous sy Neurosecretion has almost disappeared from the axons.

to moulting. At the same time, at all moults except the last, the corpora allata secrete a hormone which ensures the development of another larval instar. During the last moult, the corpora allata are inactive; and the adult is produced (Figs. 2.12, 6.4).

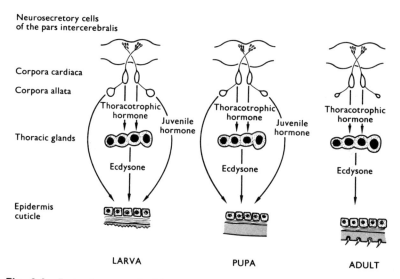

Fig. 6.4 Endocrine control of insect metamorphosis. Three hormones are involved: *thoracotrophic hormone* from the cerebral neurosecretory cells is released into the blood from the corpora cardiaca and stimulates the thoracic glands (or their equivalents) to produce *ecdysone*. Ecdysone induces the epithelial cells to begin the processes which lead to moulting. In all larval instars except the last, the corpora allata secrete *juvenile hormone* which ensures the development of another larval form. In the exopterygote (hemimetabolous) insects, progressively less juvenile hormone is secreted during each instar and results in progressive differentiation towards the adult form. In the endopterygote (holometabolous) insects, a consistently high level of juvenile hormone is secreted in each larval instar which maintains the more or less constant larval form. Juvenile hormone concentration is diminished in the last larval instar and the pupation moult results. Before the final moult in both exopterygote and endopterygote insects, the corpora allata are relatively inactive, little or no juvenile hormone is secreted, and the adult form is produced.

These hormones are named according to their function. The brain hormone, because of its effect upon the thoracic glands is called **thoracotrophic hormone**; that from the thoracic glands, **ecdysone** because it initiates moulting (ecdysis); and that from the corpora allata, **juvenile hormone (or neotenin**), because it ensures the retention of juvenile characters.

Upon what evidence are these statements based? In insects the classical extirpation: reimplantation experiments (Chapter 1) are not always possible, but the equivalent of gland removal can be obtained by ligaturing off that part of the body containing the suspected endocrine organ.

Thoracotrophic hormone

In many species of insect, the cerebral neurosecretory cells have been removed or destroyed by cautery or by x-rays.[271, 261, 133, 112] When the operation is performed early enough in the instar, the insect does not moult. Instead, it often goes into a state of suspended development for what may be a very long time, similar in many respects to the developmental arrest, or diapause, which intervenes normally in many life histories (see p. 142ff). When neurosecretory cells, or even whole brains, are reimplanted into such operated animals, their suspended development is lifted and they subsequently moult. The cerebral neurosecretory cells are therefore essential for the moulting process.

When the cerebral neurosecretory cells are removed or destroyed, neurosecretion disappears from the corpora cardiaca.[133] Ligaturing or cutting the neurosecretory nerves between the brain and the corpora cardiaca similarly depletes the amount of neurosecretion in the glands, but in addition neurosecretion accumulates on the *brain* side of the ligature or cut (Plate 5).[234, 261] This can only mean that neurosecretion is transported from the cells in the brain to the corpora cardiaca. In fact, in a very few insects, the movement of material along the living nerves has been seen.[261]

When whole brains are implanted immediately into individuals whose own neurosecretory cells have been removed, moulting is delayed compared with normal animals. Moulting is similarly delayed, but not prevented, when corpora cardiaca are removed. The reason for this is that the neurosecretory nerves, severed in the two kinds of operation, have to develop a new neurohaemal organ before release of the hormone into the haemolymph approaches normal (Plate 6).[257]

Although these experiments show the necessity of an intact cerebral neurosecretory system for normal moulting, the thoracotrophic action of the brain hormone is not yet proved. This can be demonstrated as follows.

The insect has its neurosecretory cells removed and is ligatured between the thorax and the abdomen. When a brain is implanted into the abdomen, moulting does not occur: brain hormone alone does not activate the epithelial cells. But when the brain is implanted into the thorax, this subsequently moults. By varying the level of the ligature, it can be shown that only that part of the body containing thoracic glands will moult after brain implantation (Fig. 6.5).[286, 275]

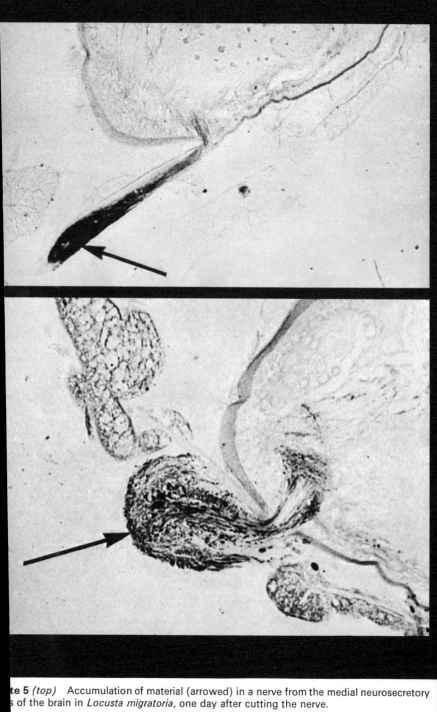

Plate 5 *(top)* Accumulation of material (arrowed) in a nerve from the medial neurosecretory cells of the brain in *Locusta migratoria*, one day after cutting the nerve.

Plate 6 *(bottom)* A new corpus cardiacum (arrowed) developed from a nerve cut (as in Plate 5) 8 days previously.

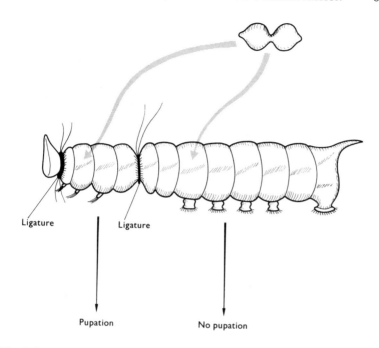

Fig. 6.5 An experiment to show the activation of the thoracic glands by the brain. A ligature behind the head isolates the source of the larva's own thoraco-trophic hormone. When a brain is implanted into the abdomen, separated by a ligature from the thorax, no pupation occurs. But when the brain is implanted into the thorax, this part will pupate due to the production of ecdysone by the thoracic glands. By varying the level of the ligature, the precise source of ecdysone can be found.

Ecdysone

The previous experiment also provides good evidence that the thoracic glands produce the moulting hormone. In many insects, definite proof by removal and reimplantation that the thoracic glands are necessary for moulting is difficult because the glands are delicate structures, almost impossible to remove completely. But the operation has been performed in a small number of insects, and has confirmed their importance for the moulting process.[256] Recently, pure ecdysone has been synthesized and is now available in large quantities. The injection of the hormone into insects ligatured to separate off their own thoracic glands is always followed by moulting.

Juvenile hormone

In many insects, when the corpora allata are removed in early instars, moulting takes place as usual, but instead of another larva being produced, a diminutive adult emerges (Fig. 6.6). When the glands are re-

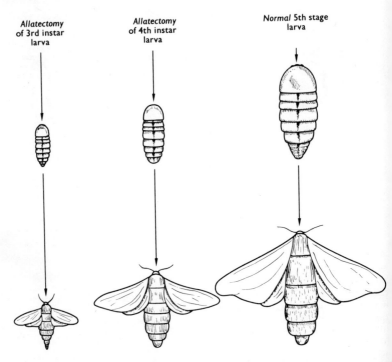

Allatectomy of 3rd instar larva   Allatectomy of 4th instar larva   *Normal* 5th stage larva

Fig. 6.6   The effects of removal of the corpora allata at different times during the life of the silkworm, *Bombyx mori*. Allatectomy during the third larval instar results in a diminutive pupa, and a correspondingly small adult. The same operation performed in the fourth larval instar produces a larger pupa and adult, although still smaller than normal. (Redrawn from Wigglesworth[277])

moved in the last larval instar, normal moulting to the adult follows. But when several corpora allata from earlier instars are implanted into the last larval instar, a supernumerary larva is produced instead of an adult. This is usually larger than any normal larva, and may subsequently moult to a giant adult. These experiments confirm that the corpora allata produce a hormone in all larval instars except the last which prevents the emergence of the adult instar.

The overall control of development by hormones thus rests upon good experimental foundation. But how is this control modified to permit the progressive differentiation towards the adult form in exopterygote insects, and the complete metamorphosis of the endopterygotes?

The classical experiments of Wigglesworth carried out upon the blood-sucking bug, *Rhodnius prolixus*, have provided an answer to this question. *Rhodnius* feeds only once in each instar, and the insect moults at a fixed time after its meal. Two or more insects can be joined together by fine capillary tubing so that their haemocoeles are in continuity[269] (Fig. 6.7); their epidermal cells grow and make contact within the insides of the tubes. Insects joined together in this way are said to be in *parabiosis*. They are like Siamese twins, with separate sets of organs, but with a common blood system. Clearly, both members of a parabiotic pair will be influenced similarly by blood borne factors. Thus the parabiotic

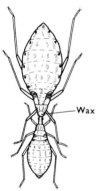

Wax

Fig. 6.7 Parabiosis between two larval stages of *Rhodnius prolixus*. One or both of the larvae are fed, parts of their heads amputated, and the individuals joined with glass capillary tubing sealed into place with paraffin wax. The elongated head of *Rhodnius* allows amputation at different levels, so that the corpus allatum can be left in place or removed as desired. (After Wigglesworth[269])

insects moult synchronously, even though only one member may have fed before parabiosis to initiate the moulting.

When a third instar *Rhodnius*, minus its corpus allatum, is joined in parabiosis to a fourth instar, which retains its corpus allatum, the fourth instar differentiates during the subsequent moult to a normal fifth instar. However, the third instar member of the parabiotic pair does not moult to a fourth instar, but under the influence of the corpus allatum of its fourth instar companion, moults to a form more like the fifth instar than the fourth (Fig. 6.8). When the contrary situation is tested—the third instar retains its corpus allatum and the parabiotic fourth instar is allatectomized—then the fourth instar moults to *another* fourth instar, instead of a fifth as it would do with its own corpus allatum[269] (Fig. 6.9). These results suggest that the third instar corpus allatum is more potent in suppressing adult differentiation than is the fourth instar gland. Similar experiments with parabiotic pairs of earlier instars lead to the conclusion that the potency of the corpus allatum in *Rhodnius* falls progressively in

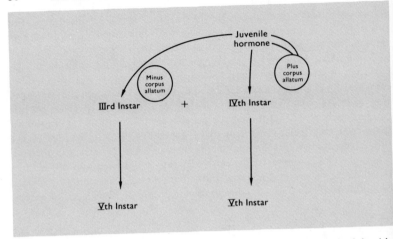

Fig. 6.8 Individuals of the third and fourth larval instars of *Rhodnius* joined in parabiosis as in Fig 6.7. The third instar is decapitated *behind* the corpus allatum, the fourth instar *in front of* the gland. Both individuals therefore develop under the influence of the fourth instar corpus allatum. The fourth instar moults normally to a fifth instar larva, but the third instar does not moult to a normal fourth. Instead, it develops more the characteristics of a fifth instar larva. Compare Fig. 6.9.

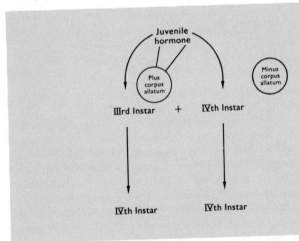

Fig. 6.9 The reverse situation of that in Fig. 6.8: the third instar retains it corpus allatum, the fourth instar is without its gland. The third instar moult normally to a fourth instar. But under the influence of the corpus allatum of th third instar larva, the fourth instar moults to a form more like another fourth insta rather than the fifth instar to which it would otherwise have given rise. Compar Fig. 6.8.

successive instars, until in the last larval instar the gland is inactive and the adult is produced. Progressive differentiation at successive moults in the hemimetabolous insects is therefore controlled by the corpus allatum.[277, 279, 284]

The results of the experiments described above could be explained in terms of a decreasing concentration of juvenile hormone in the blood— the corpus allatum becomes less 'active' in successive instars. But another explanation is possible, as the next experiment shows. Two fourth instar *Rhodnius* are joined in parabiosis, one member of the pair seven days after feeding, the other one day after feeding. The physiologically younger larva moults to a form intermediate between the fourth and fifth instars (Fig. 6.10). So when juvenile hormone is introduced *earlier*

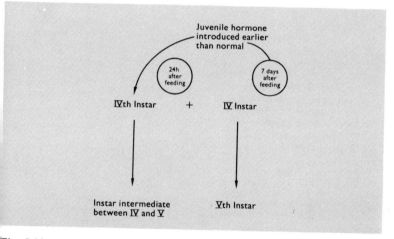

Fig. 6.10 Two fourth instar larvae of *Rhodnius* joined in parabiosis, one 7 days after feeding the other 1 day after feeding. The physiologically older individual moults normally to a fifth instar form, but the physiologically younger individual moults to a form intermediate between the fourth and fifth instars. In this individual juvenile hormone has been introduced earlier than normal in the moulting process, and the resulting larva is more juvenile than it would otherwise be. (After Wigglesworth[277])

than normal into the moulting process, differentiation towards the adult form is delayed. Thus the control over metamorphosis in hemimetabolous insects exercised by the corpus allatum can be effected in one of two ways: either by the gland becoming less active in successive instars, or by the gland secreting its hormones progressively later in successive instars. In practice, there is little difference between these two possible situations.

Does the corpus allatum control metamorphosis in holometabolous insects in a similar way? Arguing from the situation in the Hemimetabola, it might be expected that the larval corpora allata secrete juvenile hormone at a *continuously* high level in the larval instar, thus preventing metamorphosis, that the glands are less active in the last larval instar to induce pupation, and that the pupal corpora allata are inactive, producing the adult after the pupal-adult moult. Gilbert and Schneiderman have shown that this is indeed so. Having developed a bioassay for juvenile hormone (p. 224), they extracted and tested the hormone from various larval stages, pupae and adults of the giant silkmoth, *Hyalophora cecropia*. Its concentration remains high in the early larval stages, falling markedly in the last instar larva and the hormone is not detectable in the later pupal stages[110] (Fig. 6.11). Moreover, the injection of purified juvenile

Stage	Juvenile hormone activity per gram fresh weight
Unfertilized eggs	8·15
7 day embryos	7·09
1st instar larvae	7·47
5th instar larvae	0·57
Diapausing pupae (male)	1·47
Diapausing pupae (female)	1·11
2-day old developing adult (male)	0
17-day old developing adult (male)	0
20-day old developing adult (male)	13·70
22-day old newly emerged adult male	201·64
Adult male, 7 days after emergence	418·00
Adult female, 7 days after emergence	11·63

Fig. 6.11 Juvenile hormone activity during the life history of *Hyalophora cecropia*. Note the very low activity in the last (5th) instar larva, and the high activity in mature males. (Data from Gilbert and Schneiderman[110])

hormone extract into pupae often results in the formation of a *second* pupal stage after the moult[108, 288]; the injection of ecdysone into last instar larvae sometimes produces a supernumerary larval stage, proving the presence of some juvenile hormone at this time, and if the last instar larvae are allatectomized early enough in the stadium (i.e. before the corpora allata begin to secrete) an adult, or a monster intermediate between a pupa and an adult, is produced.

Moulting in adult insects

Many apterygote insects moult frequently as adults, whereas the pterygotes normally do not with the exception of the preimago of the mayfly, which moults a fine, transparent cuticle shortly after emergence. What is the cause of this difference? Simply, the thoracic glands of the pterygote insects degenerate and disappear during or shortly after the last moult, while those of the apterygote insects do not. Adult pterygotes can be induced to moult by exposing them to moulting hormone by parabiosis or implants of thoracic gland[271, 275] (Fig. 6.12). The reason for the

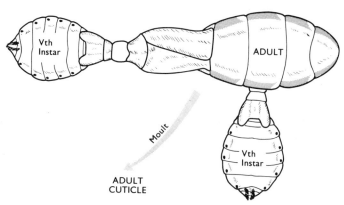

Fig. 6.12 Adult moulting in *Rhodnius prolixus*. Parabiosis between an adult and two fed fifth instar larvae causes synchronous moulting in all three individuals: the adult moults into a second adult instar. Compare Fig. 6.13. (After Wigglesworth[271])

dissolution of the thoracic glands is a little obscure: they break down during a moult which follows allatectomy, whereas if corpora allata are implanted into last instar larvae to produce supernumerary larval instars, the thoracic glands persist. In other words, the structural integrity of the thoracic glands is maintained during a moult in which juvenile hormone is present, but breaks down on undergoing a moult in the absence of the hormone.

It has been suggested that an additional factor (perhaps hormonal) is necessary for breakdown of the thoracic glands.[284, 278] But this would make the process unnecessarily complicated. In the holometabolous insects particularly, many purely larval tissues disappear during metamorphosis. The thoracic glands are larval tissues *par excellence*. There is, therefore, no need to invoke a special dissolution factor for the thoracic glands, additional to the factors which are normally involved in metamorphosis.

Mode of action of developmental hormones in insects

In the presence of ecdysone alone, the insect moults and differentiates fully to the adult form. When juvenile hormone is also present, adult development is prevented to a greater or less degree (p. 84). Ecdysone

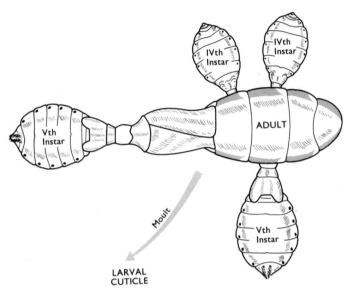

Fig. 6.13 As in Fig. 6.12, except that two fourth instar larvae, decapitated in front of the corpus allatum, are additionally joined in parabiosis to the adult. Juvenile hormone supplied by these larvae influences the moult so that the adult regresses to a more juvenile form. Legs have been omitted from individuals, and the adult wings cut short. (After Wigglesworth[271])

has consequently been called a *moulting and metamorphosis* or a *growth and differentiation* hormone, and juvenile hormone an *inhibitory* or *status quo* hormone. But the functions of the hormones implied by these names need to be examined very carefully.

There is abundant evidence that in the absence of ecdysone, its target tissues neither grow nor develop and go into a developmental arrest very similar to natural diapause (p. 83). In *Rhodnius*, exposure of the epidermal cells to ecdysone causes enlargement of the nucleolus, the appearance of RNA in the cytoplasm, and an increase in the numbers, size and branching of the mitochondria[279, 280] (Fig. 6.14). These changes in the cells are clearly those associated with the restoration of protein synthesis.

The modern view of cellular protein synthesis is that the genetic material in the nucleus—deoxyribonucleic acid (DNA)—controls the manufacture of many specific kinds of ribonucleic acids (RNAs), each RNA built on a template of DNA consisting perhaps of a temporarily unwound part of the DNA double helix. The messenger RNA (m-RNA) passes out of the nucleus and becomes associated with protein bodies called ribosomes, located along the endoplasmic reticulum of the cell where they have ready access to raw materials. Meanwhile, small molecular weight RNAs, called transfer RNAs, become bonded to free amino acids within the cell cytoplasm. The individual constitution of each transfer RNA decides which amino acid will become attached. The combined transfer RNA and amino acid passes to the ribosome, and each transfer RNA associates with a particular region of the m-RNA. Consequently, a sequence of amino acids is brought into close contact and a specific protein can be formed. The specificity of the protein is thus determined by the structure of the m-RNA on which it is formed, which in turn is dependent upon its original DNA template.

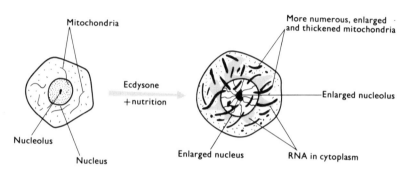

Fig. 6.14   The effects of ecdysone upon an epidermal cell in *Rhodnius prolixus*. The cell nucleus and nucleolus enlarge, RNA appears in the cytoplasm, and the mitochondria increase in size and number. These changes are associated with the restoration of protein synthesis by the cell. (After Wigglesworth[283])

Is it possible, therefore, that ecdysone can affect the nuclear DNA (the genes) in a cell so that specific proteins are produced? Karlson and his colleagues have investigated this possibility in considerable detail, working with one part of the many processes which characterize metamorphosis in the blowfly, the tanning of the puparium. Cuticular tanning is brought about by the cross bonding of protein molecules to form a stable, large molecular weight compound called sclerotin, through the intervention of a tanning quinone (Fig. 6.15). The tanning quinone in the blowfly puparium is derived from N-acetyl-dopamine. N-acetyl-dopamine comes ultimately from tyrosine; this is oxidized to dihydroxy-phenylalanine (DOPA) which is decarboxylated to dopamine. The dopamine is acetylated and oxidized to the tanning quinone: the two key enzymes in the process are dopa-decarboxylase and N-acetyl-dopamine phenoloxidase (Fig. 6.15).

Dopa-decarboxylase increases in the blowfly larva at the same time that ecdysone is produced, and also increases 7–10 hours after the injection of ecdysone into ligated larval abdomens; this increase can be inhibited by the simultaneous injection of actinomycin, mitomycin or puromycin, antibiotics which are known to inhibit RNA and protein biosynthesis. So it is concluded that ecdysone induces dopa-decarboxy-

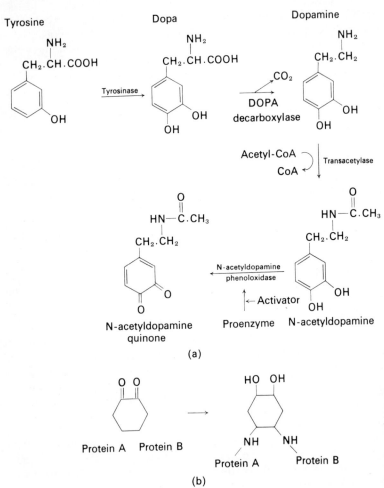

Fig. 6.15 (a) Synthesis of N-acetyl dopamine quinone in the puparium of the blowfly, *Calliphora erythrocephala*. (b) Production of sclerotin: two protein chains are linked by the quinone molecule and by repetition a complex quinone: protein macromolecule, called sclerotin, is formed.

lase by the *de novo* synthesis of enzyme protein, and RNA is needed for the process. The N-acetyl-dopamine phenoloxidase in the cuticle exists initially as a pro-enzyme; the phenoloxidase is produced as the result of the action of an activator enzyme (Fig. 6.15). Production of this activator enzyme is also controlled by ecdysone, possibly again by a process of *de novo* synthesis.[165, 166]

When ecdysone with some of its hydrogen atoms replaced by the radioactive isotope tritium is injected into last instar blowfly larvae, about 45% of the radioactivity in the epidermal cells is recovered from the nuclei, 31% from the mitochondria and 25% from the microsomes, possibly indicating a primary effect upon the nucleus. After exposure to ecdysone, isolated epidermal cell nuclei will incorporate labelled uracil into m-RNA more rapidly than nuclei which have not been so exposed. The m-RNA produced after ecdysone induction will stimulate protein synthesis in an *in vitro* system, and the protein so produced has dopa-decarboxylase activity.[167] Thus in the normal *Calliphora* larva, ecdysone is said to induce dopa-decarboxylase in the epidermal cells by activating directly the nucleus to produce a specific m-RNA. In other insects, also, DNA synthesis is said to be a primary result of ecdysone stimulation.

Supporting evidence for this point of view comes from experiments with the 'giant' or polytene chromosomes of Diptera. These polytene chromosomes consist of a large number of threads lying side by side. The chromosomes have a banded appearance, segments rich in DNA alternating with segments containing much less (Fig. 6.16). All the major landmarks of the chromosomes can be recognized in all tissues in which the chromosomes are reasonably well developed, although the actual form of the chromosomes may differ in different tissues[10] (Fig. 6.16).

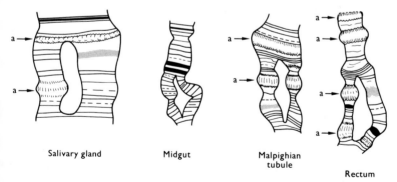

Salivary gland   Midgut   Malpighian tubule   Rectum

Fig. 6.16   Part of a giant (polytene) chromosome in different tissues of the larva of *Chironomus tentans*. The chromosome is recognised in each tissue by a small inversion. Different regions (a) are puffed in the different tissues. (After Beerman[10])

At the same developmental stage specific bands are inflated (puffed) in specific tissues. The puffs represent the main sites of RNA synthesis on the chromosomes. This chromosomal RNA differs significantly in its physical properties when extracted from different puffed segments, and the puffs consequently seem to elaborate specific m-RNAs. It is possible that the specific metabolism of different tissues is therefore controlled by m-RNAs generated in the tissue-specific puffs.[84] The non-puffing segments on the chromosomes would be suppressed genes, presumably the result of the processes of determination and differentiation in embryogenesis.

In addition to such tissue-specific puffing patterns, further series of puffs appear on the chromosomes which are related to the *developmental* stage of the insect (Fig. 6.17). Thus in the forelobe of the salivary gland

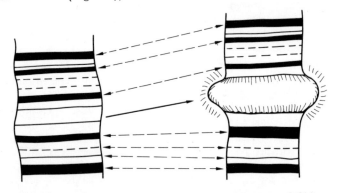

Fig. 6.17   A stage specific puff in part of a giant chromosome of *Chironomus tentans*. The same bands can be identified in both the mid-larval (on the left) and late larval (on the right) stages. One of the bands is puffed in the late larval, but not in the mid-larval, stage.

of *Acricotopus lucidus*, a very large puff (Balbiani ring) is present in the mid-larval stage and retracts at the end of larval life.[206] Such stage specific puffs have now been identified and tabulated in a number of species. Are these puffs therefore connected with the presence of hormones in particular developmental stages?

It is known that such a connexion does exist. The injection of pure ecdysone into mid-larval stages of the midge *Chironomus tentans* results in a puffing sequence in the salivary gland chromosomes characteristic of the *late* larval stage.[59] The first chromosomal puff appears only 15–30 minutes after injection, to be followed by a second 15–30 minutes later. Then 15–20 hours after the injection of ecdysone, a large number of further chromosomal puffs appear with still more forming 30–70 hours later. Among these later puffs are those characteristic of the prepupal period. If insufficient ecdysone is injected, the two very early puffs rapidly regress, and none of the subsequent puffs appear. The later puffing sequence is dependent upon the prior activity of the first two gene loci.

In the fruitfly, *Drosophila melanogaster*, the transplantation of salivary glands from younger to older larvae will induce chromosomal puffing, and such puffs are prevented from appearing when older salivary glands are transplanted to young larvae.[9] A ligature around the body which separates the salivary glands into two parts, and prohibits also the passage of endogenous hormones, prevents the appearance of chromosomal puffs in the salivary gland cells posterior to the ligature while anterior to the ligature the chromosomes puff normally. Chromosomal puffs which appear after the injection of ecdysone are therefore not experimental artifacts.

There is more direct evidence that chromosomal puffs are related to

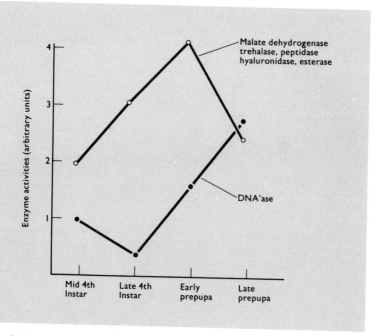

Fig. 6.18 Enzyme activities in the salivary glands of *Chironomus thummi* at different developmental stages. Only DNAase activity increases during the pre-pupal stages, associated with cellular dissolution. For further details, see text. (After Laufer and Nakase[184])

the metabolic activities of the cell. In *Chironomus tentans*, a specific type of secretion granule is produced by certain cells of the salivary glands, and there is a particular puff in one of the chromosomes of these cells. When *Chironomus tentans* is crossed with a sibling species, *Chironomus pallidivittatus*, which possesses neither puff nor secretion granule, the puff and granule are inherited in simple Mendelian fashion, and only those individuals with the puff produce the secretion.[11] However, in *Chironomus thummi*, there are only two tissue specific salivary gland puffs whereas many enzymes are produced by the glands. Since the enzymes are also present in other tissues of the body, and can be transported from the haemolymph borders of the salivary glands to their secretory canals, the relationship between the chromosomal puffs and enzyme manufacture is rather obscure. But it is possible that the salivary gland puffs in *Chironomus thummi* are in fact controlling the formation of packaging materials for the enzymes as they traverse the cells. Thus DNAase, which destroys DNA, and which is present in the form of granules in the salivary gland secretion, accumulates in the salivary glands before the onset of metamorphosis at a time when the other enzymes are decreasing in amount (Fig. 6.18). The disappearance of the tissue specific puffs at this time could perhaps result in the loss of the membranes needed to isolate this and other enzymes, so that the gross disruption of the salivary glands characteristic of metamorphosis is facilitated. The salivary gland cells would be genetically programmed for death, and the sequence would be initiated at the appropriate time by the insect's hormones.[182, 183, 184, 185]

Changes in the puffing patterns of the giant chromosomes of Diptera due to the injection of ecdysone, or to variations in the titre of the endogenous hormone, strongly supports the view that the hormone exerts its effect upon the genes. The very short time which elapses between ecdysone injection and the appearance of the first two chromosomal puffs could indicate that the hormone affects the genes *directly*. The antibiotic actinomycin blocks the DNA synthesis in cells; when this is injected together with ecdysone into *Chironomus* larvae, the first two puffs on the salivary gland chromosomes do not appear,[61] providing even more convincing evidence that ecdysone is affecting the gene loci directly. This work on *Chironomus* provides a sound basis for the conclusion that ecdysone stimulates the *de novo* synthesis of enzyme proteins during metamorphosis in *Calliphora*.

But is this really so? When the puffing patterns of the salivary chromosomes of *Chironomus* are examined throughout larval life, it is found that the early two puffs appear before *every* moult. It is only in the later sequences of puffs that differences appear between larva to larva moults and the larva to pupa moult.[60] Consequently, it must be concluded that

the reaction of the early gene loci to ecdysone is not specifically associated with metamorphosis, but with moulting in general. The hormonal
difference between the larval and metamorphosis moults lies in the
presence of a high concentration of juvenile hormone in the former and a
low or zero concentration in the latter. So it is not unreasonable to
assume that the later sequences of gene activities differ between the two
kinds of moult because of these differences in the concentration of
juvenile hormone. Therefore the *de novo* enzyme syntheses associated
with metamorphosis would be due not to the presence of ecdysone but
as a major consequence of the absence or reduced concentration of
juvenile hormone. This is to say that the genes in each cell follow a particular programme of activities during a moult in the presence of juvenile
hormone, and different programmes during moults in which juvenile
hormone is present in reduced concentration or is absent. So the major
hormone affecting differentiation in insects is the juvenile hormone and
not ecdysone. But what then is the function of ecdysone?

In *Chironomus*, if the two early chromosomal puffs are prevented from
appearing because actinomycin is also injected together with ecdysone,
or if the puffs rapidly regress because ecdysone is injected in low concentration, the later sequences of gene loci remain inactive and do not
puff.[61] So the gene activities concerned with larval or metamorphic
processes depend upon the prior activity of the genes which respond to
ecdysone. The antibiotic puromycin will block cellular protein synthesis, not at the level of DNA synthesis like actinomycin, but at the
cytoplasmic level. When puromycin is injected with ecdysone into
Chironomus larvae, the two early puffs appear as usual, but all the later
puffs are completely inhibited.[61] Consequently, the activity of these gene
loci must depend upon some interaction between the early reacting genes
and the cytoplasm. In practice, it can be said that ecdysone activates the
cell through its effect upon the initial two gene loci. Moreover, since the
subsequent gene activities depend upon interactions between the nucleus
and the cytoplasm, juvenile hormone could exert its effects at either the
level of DNA, or at some other point in the protein synthetic pathway.

Is there any evidence for this view of the importance of juvenile
hormone as a programmer of gene activities, and of ecdysone as an
activator of the cell? An adult *Rhodnius* can be induced to moult by
joining it to moulting 5th stage larvae (Fig. 6.12) or by implanting active
thoracic glands. But when juvenile hormone is also present, for example
when the adult is jointed to moulting 4th instar larvae (Fig. 6.13), the
adult moults but its cuticle reverts to a larval form. Pupal epidermis
from some insects implanted into larval hosts moults synchronously
with the larvae and produces progressively more juvenile cuticle.[224, 270]
Such reversal of metamorphosis does not occur in all insects, but that it

does in some means that juvenile hormone must play a more positive part in determining the degree of differentiation than merely by opposing its natural expression. Juvenile hormone cannot be merely an inhibitory or status quo hormone.

Insect growth is often described as discontinuous, periods of increasing size corresponding with each moult and alternating with static periods between moults. But when growth is measured as increase in wet or dry weight, or in total nitrogen, no discontinuities appear in the growth curve except for small losses associated with ecdysis (Fig. 6.19). Although

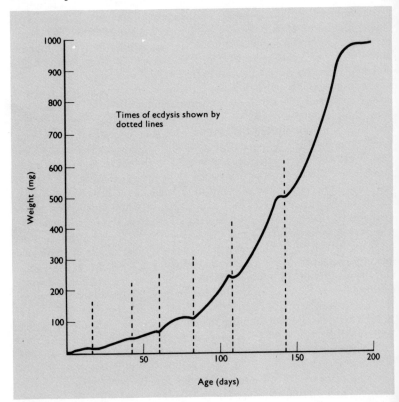

Fig. 6.19 Growth in the stick insect, *Carausius morosus*. Growth is a continuous process during each instar, the small discontinuities at each ecdysis being due to loss of the cast cuticle.

the production of ecdysone is cyclic, the growth increments continue throughout each larval instar, and even in the adult when the thoracic glands have disappeared.[147] Individual tissues in the adult show the same kind of growth increments so the weight increases are not associated solely with the storage of reserves (Fig. 6.20). Moreover, some insects can

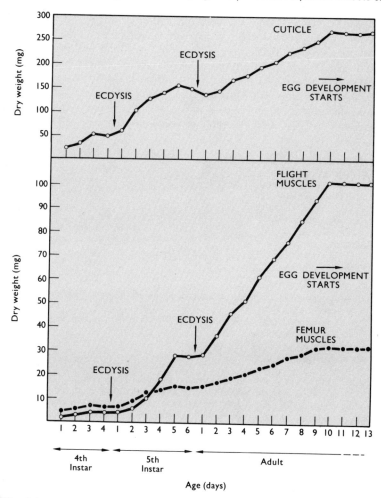

Fig. 6.20   Growth of the cuticle, flight muscles and femoral muscles during the fourth and fifth larval and the adult instars of the desert locust. The tissues increase in weight not only during the larval instars, but also in the adult after the final moult. Somatic growth in the adult female continues until just before oocyte development begins. (After Hill and Goldsworthy[146]; Hill, Luntz and Steele[147])

be induced to moult without any associated increase in size.[197] Neither is cell division causally related to the action of ecdysone. The imaginal buds of many holometabolous insects increase in size many times by cell division (Fig. 6.2), and can undergo radical changes in form between larval moults. In larvae of *Locusta*, two waves of cell division pass forwards and backwards from the posterior and anterior ends of the body before the moult, related to but clearly not dependent upon the presence of ecdysone.[256] Cell division takes place during wound healing, even in adult insects where the prothoracic glands are no longer present. In *Rhodnius*, cell division in the epidermis and fat-body is induced by the mutual separation of the cells resulting from the stretching of the abdomen after feeding, and mitosis in the fat-body occurs in the absence of ecdysone[283] (Fig. 6.21). Greater or less mitotic activity in different

Tissue	% increase in nuclei/unit area
Epidermis : stretched* by feeding +MH and JH	110
Epidermis : unstretched** +MH and JH	4
Fatbody : stretched*** +MH	89
Fatbody : unstretched** nutrition +MH	0
Fatbody : stretched**** nutrition alone	50

* 4th stage larva fed, decapitated after 24 hours and joined in parabiosis with 4th stage larva.
** 4th stage larva *unfed*, decapitated after 24 hours and joined in parabiosis with 4th stage larva.
*** Normal 4th stage larva after feeding.
**** 4th stage larva fed, decapitated 24 hours later.

Fig. 6.21   Mitosis in abdominal epidermal cells and fatbody cells during moulting in *Rhodnius prolixus*. In the epidermis, the increase in number of cells depends more upon stretching than upon the presence of hormones. Increase in cell number in the fatbody depends wholly upon stretching. (After Wigglesworth[283])

parts of the body depends upon whether or not juvenile hormone is present.[271, 272]

These observations suggest that individual cells have an inherent capacity for the kind of growth that they will undergo during any moult. This conclusion is underlined by the demonstration that every epidermal cell in *Rhodnius* has the capacity to divide and differentiate to form sensory plaques and dermal glands, and the expression of this potentiality is not dependent upon the presence of ecdysone or even juvenile hormone.[276, 277] Further, the overall pattern of pigmentation, areas of sensory plaques, hairs and even cuticular structure are fairly constant for any larval instar, and will vary predictably at each moult until the final adult pattern is attained. In *Rhodnius,* when pigmented epidermal cells migrate to heal a wound in the larval stages, the degree of displacement can be traced through each successive instar to the adult (Fig. 6.22): the

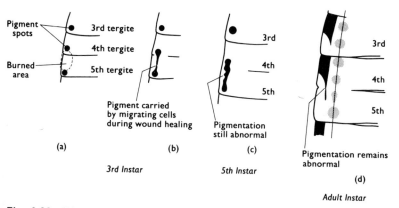

Fig. 6.22   Wound repair and cell lineage in *Rhodnius prolixus.* When a small part of the tergum of a third instar larva is burned (a), the area is healed by division and migration of the surrounding cells. If these cells are pigmented, the normal pigment pattern becomes distorted (b). This distortion is still evident in successive instars (c), and even in the adult instar (d), where the relationship between pigmented and unpigmented cells is different from that in the larvae. This experiment suggests the continuity of a cell line throughout development in this exopterygote insect. Compare Fig. 6.3. (After Wigglesworth[272])

cells retain the properties they had in their original site.[271, 272] So even in the larval stages, the epidermal cells possess the latent potentialities for particular adult structures. Thus the larval cells reveal characteristics which typify the larval body, and the same cells produce the characteristics of the unique body form of the adult of any particular species. In other words, the larval cells carry a dual pattern: a visible larval pattern, and a latent, invisible adult pattern.[271] The juvenile hormone can thus maintain the larval pattern perhaps indirectly by participation in cytoplasmic processes, or even perhaps by some action at the level of the genes.[277]

   This interpretation of the events associated with metamorphosis recognizes the overriding importance of the individual cell in the determination of its specific characters. Cell division and differentiation are the intrinsic properties of the cell, residing in its genes, the particular expression of which at any stage is controlled by the juvenile hormone. Although ecdysone induces the activation of its target cells, such as the epidermis, which is necessary before growth and differentiation can occur, other cells may be activated by quite different factors. Thus the fatbody cells of *Rhodnius* initiate protein synthesis when nutrients in the haemolymph become available after a meal.[283] Even so, their subsequent metabolism can be modified by circulating hormones, as when specific vitellogenic proteins are produced in the adult female (Chapter 7).

Some species of lepidopteran pupae, and adults, retract the epidermis from the cuticle when injected with ecdysone and mitomycin, but do not secrete a new cuticle or show any other signs of further development.[200] Mitomycin, like actinomycin, blocks DNA synthesis, and the subsequent lack of cellular activity after its administration is not surprising. But the separation of the epidermis from the cuticle, called apolysis, is normally a very early step in the moulting process and would suggest that the cells have been activated by ecdysone even though DNA synthesis is inhibited by mitomycin. This result would seem to contradict that in *Chironomus*, where the injection of ecdysone and actinomycin prevents the appearance of the early puffs on the salivary gland chromosomes (p. 97). It is possible, of course, that in both the lepidopterans and the chironomids, ecdysone has a primary effect upon some cytoplasmic constituent(s) with a consequent effect upon the genes, and that apolysis is the result of this primary effect within the epidermal cell. But the actions of antibiotics upon cellular protein synthesis in different animals are not necessarily the same, nor are the effects of the same hormone upon *dissimilar* tissues in *different* animals. Moreover, the doses of ecdysone used in the experiments upon the moths were rather large, so that their effects could possibly be abnormal.

7

Endocrine Mechanisms in the Insecta—II

Hormones and Reproduction

THE INSECT REPRODUCTIVE SYSTEM

Insects generally are bisexual, and although parthenogenesis is anything but uncommon, frequently playing a very important part in normal life-cycles (e.g. those of many aphids), sexual reproduction is the usual rule.

Until recently, it was considered that the morphological differences between male and female insects were strictly genetically determined. The occasional appearance of intersexual mosaics provided good evidence for this view. It was thought that the development of secondary sexual characteristics, controlled by hormones produced by the gonads, so prominent a feature of sexual differentiation in the vertebrates, just did not apply in insects. But it is now known that in one insect at least (the glow-worm, *Lampyris noctiluca*) both the primary and secondary male characteristics are induced by an 'androgenic' hormone from paired endocrine glands associated with the testes[211] (Fig. 7.1). Transplantation of these glands into a female larva at a fairly early stage will induce masculinization. It is too early yet to say how widespread in insects this phenomenon may be, but even this one example shows the danger of generalizing about such aspects of the life of insects.

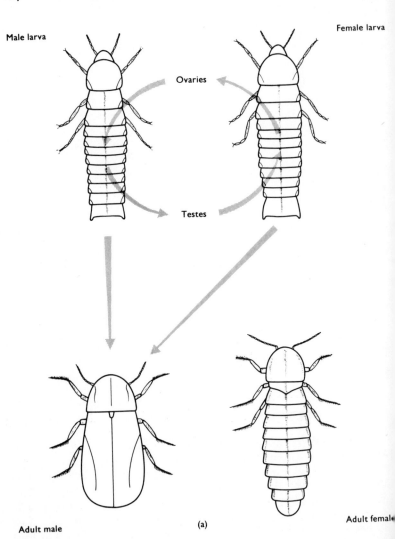

Fig. 7.1 Sex determination in *Lampyris noctiluca*. (a) The transplantation o testes together with androgenic tissue from a male to a female larva causes the latter to develop into a male. Ovary transplantation from female to male larvae i without effect. (b) *opposite* The table shows that the potency of testes to induce masculinization of females is reduced in the pupa and is negligible in the adul Correspondingly, the competence of the female to react to the factor is lost afte pupation. (After Naisse[211])

Testes from	Implanted into	Result
male larvae between moults 3 and 4	female larvae of same age	all males produced
	female larvae between moults 4 and 7	all males produced
	female pupae	females produced
male larvae between moults 4 and 5	female larvae between moults 3 and 4	all males produced
	female larvae of same age	all males produced
	female pupae	females produced
male pupae	female larvae between moults 3 and 4	80% males produced
male adults	females of all stages	all females produced

(b)

Fig. 7.1 Continued

The terminal segments of the abdomen in both male and female insects are deeply involved in the formation of the genitalia (Fig. 7.2). In the female, various kinds of ovipositor are developed from what must have been originally abdominal appendages; in the male, outgrowths are formed from the terminal segment which divide to form a protrusible phallus and associated clasping organs. In both sexes in addition, parts of the internal ducts and glands of the reproductive system are derived from embryonic ectodermal invaginations. The development of adult genitalia is a major characteristic of insect metamorphosis. In a limited sense, therefore, the combination of hormones which controls metamorphosis (p. 81) may also be considered to influence reproduction. But in general, reproductive endocrinology confines itself to those hormones which are active in the adult instar itself.

In most insects the ovaries are paired, each surrounded by a thin epithelial sheath containing connective tissue fibres and unstriped muscle cells in addition to the flattened epithelial cells. Each ovary is made up of a number of egg-tubes, or ovarioles, the proximal extensions of which are called the terminal filaments and attach the ovaries to the dorsal body wall (Fig. 7.3). There are usually 4, 6 or 8 ovarioles in each ovary, although numbers varying from one to several hundred are not uncommon. In queen termites, there may be several thousand ovarioles in each ovary.

Beneath the terminal filament in each ovariole is the germarium, in which the differentiation of oocytes from the primary germ layer takes place. Next in line along the ovariole is the vitellarium, which makes up the greater part of each ovariole and in which yolk deposition or vitellogenesis in the oocyte occurs. The ovariole joins the oviduct by a narrow neck or pedicel.

Two main types of ovariole are found in the insects, separated according to the way in which the oocyte is differentiated from the oogonium and the way in which it is nourished during the early growth period. In the *panoistic ovariole*, the daughter cells of the oogonia metamorphose into oocytes (Fig. 7.6). These become associated with prefollicular cells

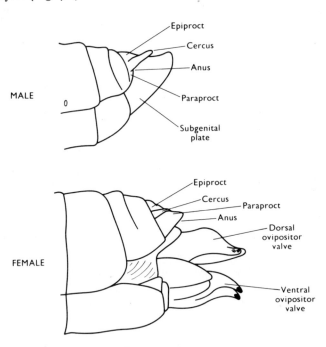

Fig. 7.2   External genitalia of the male and female desert locust, *Schistocerca gregaria*. The development of these specifically adult characteristics is controlled by the juvenile hormone during moulting.

of mesodermal origin, a little lower down the germarium. The prefollicular cells become organized into a layer of follicle cells surrounding each oocyte and accompany the latter into the vitellarium. The follicle cells are concerned with all the stages of growth and yolk deposition in the oocyte, and after vitellogenesis is completed, the follicle cells secrete the shell, or chorion, around the egg. In the *meroistic ovariole*, on the other hand, only one of the daughter cells from each oogonium metamorphoses into an oocyte (Figs. 7.4, 7.5). The remaining daughter cells are connected with the oocyte by shorter or longer intercellular connections and nourish the oocyte during the early stages of its growth, or rather, supply it with material for biosynthesis; these associated cells are consequently called nurse cells. The meroistic ovariole may be further

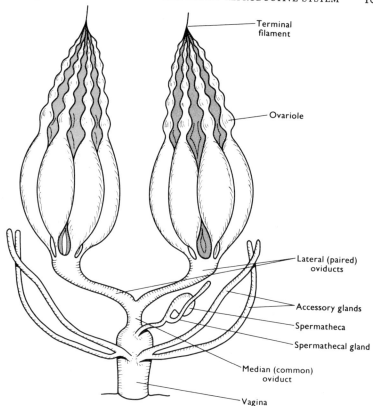

Terminal filament

Ovariole

Lateral (paired) oviducts

Accessory glands

Spermatheca

Spermathecal gland

Median (common) oviduct

Vagina

Fig. 7.3. Ovaries of a generalized insect. Each ovary is composed of a number of egg tubes, or ovarioles, opening into a lateral oviduct. The paired oviducts lead into a common oviduct, from which a spermatheca for sperm storage opens. A spermathecal gland may be present, whose secretion may keep the sperm viable, or aid their movement down the spermathecal duct during fertilization. Accessory glands produce secretions to surround the eggs and lubricate their passage during oviposition.

subdivided into two kinds. In one, the nurse cells remain in the germarium, and as the oocyte moves into the vitellarium its connections with these cells become extended to form nutritive cords (Fig. 7.5). In the other, the nurse cells remain in contact with the oocyte, moving with it into the vitellarium (Fig. 7.4). The first kind of ovariole is called **telotrophic** (feeding from afar) and the second, **polytrophic**. Prefollicular cells are associated with the oocyte in both kinds of meroistic ovariole: in the telotrophic ovariole, only the oocyte becomes invested with follicle cells, which take over the nourishment of the oocyte when the nurse cells degenerate; in the polytrophic ovariole both nurse cells and oocyte are invested with follicle cells, although there is often a pronounced constriction between the two. The nurse cells degenerate at a fairly early stage in the development of the oocyte, but their remains can persist until a much later stage in vitellogenesis. The follicle cells in the meroistic ovaries finally produce the chorion of the egg in the same way as in the panoistic ovariole.

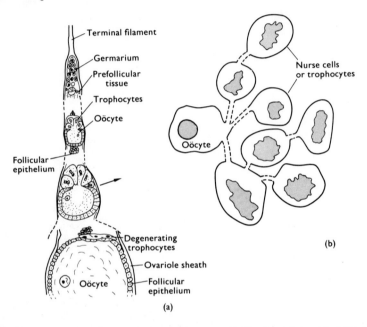

Fig. 7.4   (a) A polytrophic meroistic ovariole. Unlike the telotrophic type, the trophocytes remain in close association with the oocyte during its early passage along the ovariole. Trophocytes and oocyte together make up a follicular chamber. (b) This shows how the trophocytes are connected with each other and ultimately with the oocyte which they supply with materials during its early development. Oocytes and trophocytes are developed from a primary oogonium.

The essential difference between panoistic and meroistic ovarioles is that in the former the follicle cells are involved in the early growth of the oocyte, whereas in the latter this primary function has been taken over by the nurse cells, themselves very much modified oocytes. In both types of ovariole, the later development of the oocyte and the formation of the chorion are functions of the follicle cells. Since vitellogenesis in successive oocytes takes place in sequence down the length of the vitellarium,

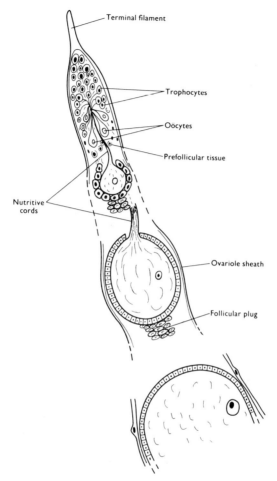

Fig. 7.5 A telotrophic meroistic ovariole. Nurse cells, or trophocytes, which have developed from primary oogonia, remain in the germarium but are attached to each oocyte by cytoplasmic filaments, the nutritive cords. The initial development of the oocytes depends upon the transmission of materials from the trophocytes along the nutritive cords.

various stages in oocyte development may be seen at one time in any kind of ovariole.

Panoistic ovarioles are more typical of the 'older' orders of insects—grasshoppers, termites, stone-flies and dragon flies, for example. The desert locust, *Schistocerca gregaria*, has panoistic ovarioles. Telotrophic ovarioles are characteristic of the bugs, including *Rhodnius prolixus*, and some beetles. Most holometabolous insects, including the blowfly *Calliphora*, have polytrophic ovarioles.

The pedicel of each ovariole leads into the mesodermal paired oviduct on its own side of the body. The paired oviducts lead into the ectodermal common oviduct, which usually opens on the ventral surface of the ninth

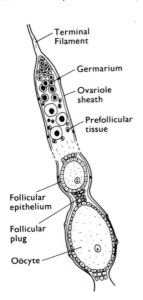

Terminal
Filament

Germarium

Ovariole
sheath

Prefollicular
tissue

Follicular
epithelium

Follicular
plug

Oöcyte

Fig. 7.6 Section through a panoistic ovariole. The oocytes are each surrounded by follicle cells shortly after their emergence from the germarium.

abdominal segment. Associated with both the mesodermal paired oviducts and the ectodermal common oviduct are the accessory glands, which are variously concerned with secreting material to protect the eggs, or to cement the eggs to an appropriate substrate, or both, after laying (Fig. 7.7).

The testes are usually paired structures in the insects (Fig. 7.8), although many species have a single, fused median organ. Each testis is enclosed within a connective tissue sheath, which may be attached dorsally to the wall of the abdomen. The most distal part of the testis is the germarium where spermatogonia are produced. These are set free in the germarium where they become encapsulated by groups of somatic cells. The transformation of the cysts of spermatogonia to spermatocytes and spermatids, and the final metamorphosis to spermatozoa, may occur in different zones in the testis distally from the germarium. The process is not unlike oogenesis in the female; much modified, of course, because there is no vitellogenesis.

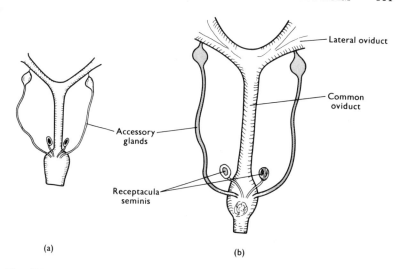

Fig. 7.7 Development of female accessory glands in the blowfly, *Calliphora erythrocephala*. (a) Reproductive system of a newly moulted female. (b) Reproductive system of a mature female.

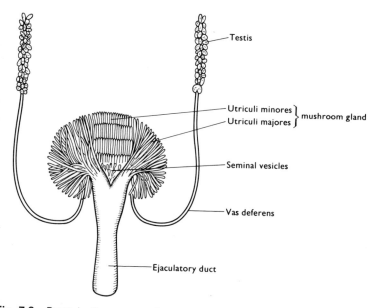

Fig. 7.8 Reproductive system of a male cockroach. The accessory glands are commonly called the mushroom gland.

The testes lead into the paired vasa deferentia, and the packets of spermatozoa are usually stored in dilatations of each vas deferens, the seminal vesicles. The paired vasa deferentia lead into a common ejaculatory duct, which opens into the phallos, and may have various structures developed within it to pump the sperm out of the system.[73] The ejaculatory duct and phallos are ectodermal in origin compared with the mesodermal vasa deferentia. Associated with both the mesodermal and ectodermal parts of the reproductive system are accessory glands, of a wide variety of form and structure in different insects, and producing a range of secretions which help in sperm transfer from the male to the female (Fig. 7.8).

All the winged insects copulate, sperm being passed from the male genitalia to the female reproductive opening during coupling. The only exception is found in the dragonflies, where the male has accessory reproductive mechanisms on the ventral side of the *anterior* abdominal segments, and sperm is transferred to these before coupling takes place. The terminal abdominal segments in the male bear clasping organs which hold the female's head during copulation, while she bends her abdomen forwards so that her reproductive opening comes into contact with the anterior accessory genitalia of the male. Sperm transfer is thus indirect in the dragonflies, a unique device in the pterygote insects, but a phenomenon not unknown in other animal groups such as the spiders, squids etc. In all other Pterygota, the direct transfer of sperm is of two kinds. In one, the sperm are passed immediately to special reservoirs— the spermathecae—inside the female by means of a very highly developed male intromittent organ. In the other the sperm are enclosed within a special container, the spermatophore, which is placed inside the female's genital opening, the sperm later passing from the spermatophore to the storage reservoirs in the reproductive tract of the female. The materials for spermatophore production are secreted by the accessory glands of the male reproductive system. The liberation of the sperm packets from the spermatophore within the female reproductive system may be the result of the digestion of the materials of the spermatophore by special secretions produced in the female tract. Often, the spermatophore is so constructed that the sperm packets are ejected from the spermatophore with some force.[68, 69] The sperm packets move up the oviducts to the spermathecae usually as a result of muscular contractions of the ducts; these contractions may be initiated by a further secretion from the male.

ENDOCRINE CONTROL OF OOCYTE DEVELOPMENT

In adult pterygote insects, the thoracic glands disappear during or soor after the last moult (p. 89). Consequently, of the organized endocrine system only the cerebral neurosecretory system and the corpora allata remain as possible sources of hormones controlling reproductive development.

In most insects, the corpora allata are essential for the full development of the eggs.[135] In the adult *Rhodnius*, a blood meal fully activates the corpus allatum and the eggs develop; development is also normal if *Rhodnius* is decapitated after the meal in front of the corpus allatum, but if it is decapitated *behind* the gland the eggs only develop to the stage at which yolk deposition (vitellogenesis) begins, when the nurse cells have completed supplying the oocytes with the materials for protein synthesis (p. 106).[269] The follicle cells then invade the oocyte and resorb it, a process known as oosorption. If the decapitated female is joined in parabiosis with a fed male, or with a female with intact corpus allatum, or if active corpora allata are implanted into the headless female, then normal oocyte development is restored (Fig. 7.9). These experiments allow three very important conclusions to be drawn:

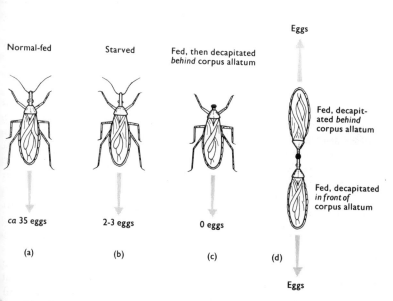

Eggs

Normal-fed   Starved   Fed, then decapitated
                                     behind corpus allatum

Fed, decapitated *behind* corpus allatum

Fed, decapitated *in front of* corpus allatum

ca 35 eggs   2-3 eggs   0 eggs

(a)   (b)   (c)   (d)

Eggs

Fig. 7.9 Experiments to show the necessity of the corpus allatum for egg production in *Rhodnius prolixus*. A normal fed female (a) produces about 35 eggs. If starved (b) a very small number of eggs is produced depending upon the amount of food present in the gut. Fed, and then decapitated behind the corpus allatum (c), no eggs are produced. Fed, and joined in parabiosis with a fed female possessing its corpus allatum, both individuals produce eggs (d). (After Wigglesworth[269])

1 The corpus allatum is necessary for egg development.
2 The corpus allatum hormone is *not* sex specific.
3 In *Rhodnius*, the neurosecretory system is not necessary for oocyte development (decapitation in front of the corpus allatum removes the brain).

In *Rhodnius*, an adult corpus allatum transplanted into an allatecto-mized larva has morphogenetic effects identical with those of the larval gland; larval corpora allata implanted into an allatectomized adult restore normal egg development. It is thus very likely that the adult gonadotrophic hormone from the corpus allatum is identical with juvenile hormone.

Similar results have been recorded from many insects from most of the major orders. The exceptions are the stick-insects (Phasmidae) and many butterflies and moths (Lepidoptera). In these groups, egg develop-ment proceeds normally in the absence of the corpora allata. The stick-insects are unusual species, normally considered to be neotenous, i.e. the reproductive individuals are *juvenile* forms so far as their somatic structure is concerned. It may not be surprising, therefore, that endocrine control over reproduction has been lost. Many Lepidoptera are short-lived as adults, implying the need for rapid maturation. In fact, many female moths emerge with almost fully developed eggs: the loss of endocrine control of maturation in these forms is also not unexpected (and see p. 233).[134]

In insects with panoistic ovarioles, allatectomy after vitellogenesis has begun results in oosorption, as in *Rhodnius* and other insects with mero-istic ovarioles.[142, 143] But allatectomy before the beginning of vitello-genesis does not cause the death and resorption of the oocytes: instead they remain quiescent, and will initiate development again if corpora allata are reimplanted. It seems that once the cellular apparatus for vitellogenesis has been established, either *de novo* in the oocytes of panoistic ovarioles, or by passage from the nurse cells in meroistic ovarioles, then development *must* proceed along its normal path. Any factor which prevents such development causes oocyte death and oosorption.

In starved *Rhodnius* adults, and in those which have been decapitated behind the corpus allatum, the oocytes develop normally up to the stage at which they lose their connections with the nurse cells, and it would seem that the vitellogenic process itself is under the immediate control of the hormone. But is the endocrine control by the corpus allatum of vitellogenesis direct or indirect? In other words, can the corpus allatum hormone in the adult be considered a gonadotrophic hormone, directly affecting yolk deposition by the developing oocytes? Or is the vitellogen promoting action indirect, some other tissue or organ being influenced by the hormone, and the effect upon vitellogenesis itself being secondary to this other action?

From the results of parabiosis between an adult female *Rhodnius minus* its corpus allatum and a fourth stage larva *plus* its corpus allatum, Wigglesworth pointed out that 'in those experiments in which no egg development occurred, the characters produced by the moulting fourth stage nymph were those of the normal fifth stage nymph. Whereas when active development of eggs took place the characters of the nymphs showed a certain degree of differentiation towards the adult form. This result could be interpreted to mean that the developing eggs were absorbing juvenile hormone and so depriving the tissues of the moulting fourth stage nymph of the supplies they need.'[272] In other words, the corpus allatum hormone of the adult female is a gonadotrophic hormone in the strict sense of the term. But this conclusion is admittedly very tentative and elsewhere Wigglesworth says: 'Associated with the development of eggs in the adult female there is much more rapid digestion of the intestinal contents. Whether this is a direct effect of the hormone on digestion and metabolism . . . or whether it is an indirect effect consequent upon the demands of the developing ovaries, has not been determined.'[269] Here the possibility is admitted that the corpus allatum hormone may not strictly be gonadotrophic, but may perhaps be a metabolic hormone, and that the ovaries develop as a result of the nutrients which are thereby made available.

Blood pigments from the host can circulate in the haemolymph of *Rhodnius* as kathaemoglobin, and can be transferred to the eggs in this form.[273] In *Rhodnius*, therefore, fairly large molecules are able to pass through the membranes of the gut and those of the follicle cells and developing oocytes. A large percentage of the yolk in the oocytes consists of protein. Are these proteins synthesized by the follicle cells and oocytes from amino acid precursors in the haemolymph? Or are they synthesized elsewhere, perhaps under the influence of the corpus allatum hormone, and transferred, like the kathaemoglobin, *in toto* to the developing eggs? And may such a transfer also be controlled by the corpus allatum hormone? In effect, is the corpus allatum hormone a metabolic hormone, or a gonadotrophic hormone, or even both?

It will be seen later that a great deal of confusion exists as to the exact role of the corpus allatum hormone in adult insects generally. But in *Rhodnius*, some attempt has very recently been made to determine the gonadotrophic or metabolic function of the hormone.

Because of differences in their constitution, particularly in their possession of different proportions of polar groups, proteins in suitable solutions and on particular substrates will migrate at different rates

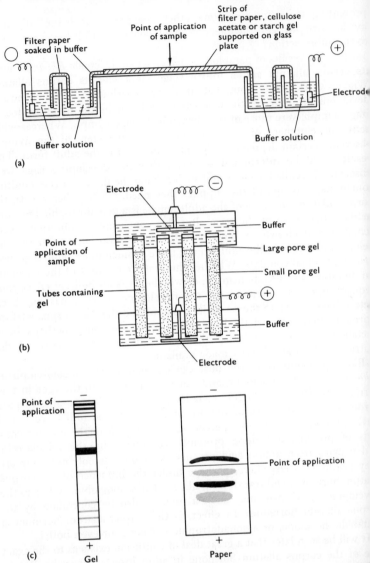

Fig. 7.10 The technique of electrophoresis. (a) An apparatus for separat~~ing~~ proteins on strips of filter paper, cellulose acetate or starch gel. (b) An appara~~tus~~ for disc electrophoresis where proteins are separated in acrylamide gel contai~~ned~~ in tubes. (c) A comparison of the separation of the blood proteins of the fer~~male?~~ desert locust on filter paper and in acrylamide gel. The gel technique greatly ~~im~~ proves the separation of the proteins.

when subjected to a voltage difference. This movement, which can be towards either the positive or the negative electrode, is called protein *electrophoresis* (Fig. 7.10). The substrates most used are paper, starch gel or cellulose acetate. The solvents are buffered to a pH which is kept constant for any series of determinations. The sample to be analyzed is placed on the substrate and the potential difference applied for a specific time. The different proteins, or groups of proteins move at their respective rates along the substrate during this time. After fixing, the protein bands can be stained, and their positions relative to the line of origin determined.

When samples of blood from last instar *Rhodnius* larvae are analyzed electrophoretically, a characteristic sequence of proteins, or protein groups, is obtained. When blood from adult females is analyzed in the same way, a similar pattern of proteins is obtained, except that two additional slowly moving bands are present.[62] Blood from adult males has the same electrophoretic pattern as that from the adult females. When the egg proteins are examined electrophoretically, fewer bands are present, but those which occur in largest amounts are the two which are additionally present in adult compared with larval blood. It is therefore very likely that it is these two proteins, or protein groups, found in the blood of the adult which are incorporated into the yolk of the eggs. (Fig. 7.11).

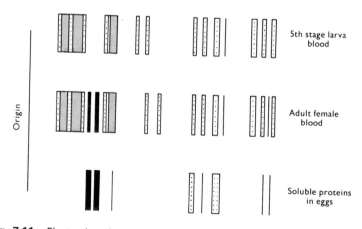

Origin

5th stage larva
blood

Adult female
blood

Soluble proteins
in eggs

Fig. 7.11 Electrophoretic separation of the proteins of the blood and eggs of *Rhodnius prolixus*. The blood of the adult female contains two protein fractions which are not present in the blood of the fifth stage larva. These same two protein fractions constitute the larger part of the soluble proteins of the eggs. This suggests that adult specific proteins appear in the blood of the adult female and are taken up by the developing eggs. (After Coles[62])

The proteins of the adult blood can be combined with a compound, fluorescein iso-cyanate, which fluoresces strongly in ultra-violet light. When this 'labelled' blood is injected into the body of an adult female, the eggs become strongly fluorescent within 24 hours. Blood from a fifth instar larva, similarly labelled and injected into the body of an adult female, results in very little fluorescence in the eggs. This suggests strongly that the adult-specific blood proteins are passed into the eggs in toto.[62]

But where are the blood proteins produced? And where does the corpus allatum hormone act in promoting the process?

When amino acids, with one of their carbon atoms replaced with the radioactive isotope C^{14}, are injected into a female Rhodnius, radioactive proteins are formed in the fatbody, and these subsequently appear in the blood. The fatbody is therefore very likely the site of synthesis of the adult blood proteins.[63]

When adult female Rhodnius are fed and then decapitated behind the corpus allatum, no yolk deposition occurs, of course, in the oocytes. But in addition, only very small quantities of the adult-specific proteins appear in the blood. When the ovaries are removed, the adult-specific proteins reach a very high concentration in the blood. Similarly, no yolk proteins can be found in the fatbody after decapitation behind the corpus allatum. Amino acids are present in high concentration in the blood after decapitation; this means that digestion of the meal is not interrupted by the operation. Finally, when the developing oocytes are incubated with fluorescein-labelled proteins in vitro, there is some uptake of protein by the oocytes, but not as much as that taken up after the labelled protein is injected in vivo.

It is concluded from these experiments that the corpus allatum in the female Rhodnius directly controls the synthesis by the fatbody of specific proteins which are passed to the developing oocytes to be incorporated into the yolk. In other words, the corpus allatum hormone in the female Rhodnius is a metabolic hormone. It perhaps acts upon specific genes in the nuclei of the cells of the fatbody, causing these eventually to produce enzymes for the synthesis of the adult specific proteins. This action of the corpus allatum hormone in the adult is therefore thought to be similar to the way in which juvenile hormone acts during growth and development (p. 97). In addition, the difference in uptake of labelled protein between oocytes in vitro and in vivo may very possibly suggest a further gonadotrophic effect of the hormone. But this evidence is not nearly substantial enough for this hypothesis to be anything but extremely tentative.

In the adult female Rhodnius, the cerebral neurosecretory system apparently plays no part in the control of egg development (p. 114).[2]

But in many other insects, removal or destruction of the medial neuro-secretory cells prevents oocyte growth and vitellogenesis. In the blowfly, *Calliphora*, and in locusts and grasshoppers, the size of the egg chambers or oocytes is much less after destruction of the neurosecretory cells than after allatectomy,[261, 133] (Fig. 7.12). In these insects, the corpora allata

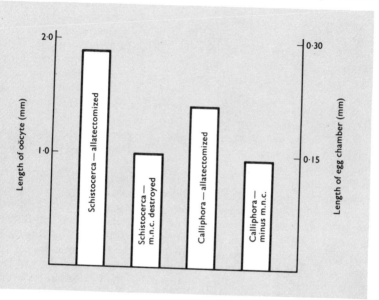

Fig. 7.12 Differential effects of allatectomy and cautery or removal of the cerebral neurosecretory cells in the desert locust, *Schistocerca gregaria*, and the blowfly, *Calliphora erythrocephala*. In both insects, the oocytes or egg chambers develop further after allatectomy than they do after destruction or removal of the cerebral neurosecretory cells. (After Thomsen[261], and Highnam[133])

are smaller after the neurosecretory cells have been removed, so part of the effect upon the oocytes could be due to these glands being inactive. But in *Calliphora*, the implantation of active corpora allata into females with their neurosecretory cells removed does not restore full growth and vitellogenesis to the oocytes. The neurosecretory cells must produce a factor affecting oocyte development in addition to any effect upon the corpora allata.

In *Calliphora*, which has polytrophic meroistic ovarioles, oocyte growth up to the stage of vitellogenesis involves the synthesis of structural pro-teins and the establishment of enzymic mechanisms derived from the nurse cells for future vitellogenesis. Since this is the stage most affected by removal of the neurosecretory cells, it is very likely that the cells in some way control protein synthesis, since enzymes themselves are of course proteins. How can this hypothesis be tested?

It is well known that blowflies will live for long periods when fed carbohydrate alone, but will not lay eggs. In what way is egg development affected in the absence of protein? This question can be answered very simply. When newly emerged flies are fed upon sugar and water, growth in the oocytes and accessory glands is retarded, and the corpus allatum fails to increase in size. These effects are remarkably similar to those which follow removal of the median neurosecretory cells. These observations provide strong support for the hypothesis that the neurosecretory cells are involved in some aspect of protein metabolism in the body.

If the neurosecretory cells are involved in some basic process concerned with protein metabolism, then it is likely that this would be reflected in the digestion of meat, the absorption of the amino acids produced by digestion, or in the resynthesis of the absorbed amino acids into yolk-proteins.

The protease activity of the gut can be estimated fairly simply by the following method. An artificial substrate is made of a protein such as casein to which is combined an azo dye. The chromophoric protein so formed is incubated with the enzyme extract from the gut, suitably buffered to the correct pH, for a fixed time. The protein is digested by the enzyme, and the decombined dye is liberated into the incubation medium. If more enzyme is present, the protein is digested more rapidly, and more dye is liberated in a given time. At the end of the incubation period, the undigested coloured protein is precipitated and filtered from the incubation medium. The concentration of liberated dye can then be measured most easily by an optical density method. Fig. 7.13 gives the relation between optical density and enzyme concentration in a typical experiment.

There is a clear cycle of protease activity in the gut of normal *Calliphora* females, which is considerably reduced when meat is omitted from the diet (Fig. 7.14). When the median neurosecretory cells are removed, and then 3 corpora cardiaca-allata implanted, the production of midgut proteases is raised almost to the normal level[263] (Fig. 7.14).

These results show: (i) that in *Calliphora* the neurosecretory system controls the production of midgut proteases, (ii) that the absence of meat in the diet reduces protease production, and (iii) that allatectomy reduces midgut protease production.

The results summarized in (ii) and (iii) could either be due to the *direct* effect of meat eating, and corpus allatum hormone, upon protease production, or else the effects could be *indirect*, mediated through the neurosecretory system. Some very detailed and carefully documented

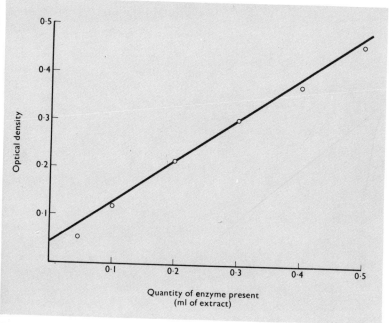

Fig. 7.13 Measurement of protease activity by the azo-casein method. The protease digests the chromophoric protein azo-casein, releasing dye into the incubation medium. The graph shows the relationship between the quantity of dye released (measured by its optical density) and the quantity of enzyme present. Increasing amounts of enzyme cause a linear increase in the amount of dye released. (After Thomsen and Moller[263])

observations on the histology of the neurosecretory system in *Calliphora* help in choosing between these alternatives.

In the newly emerged female fly, the neurosecretory cells contain very little intracellular material and their nuclei are large (Fig. 7.15). When the flies are fed only sugar, the cells become stuffed with material and their nuclei are reduced in size[264] (Fig. 7.15). The most reasonable explanation for these observations is that in the newly emerged flies, the neurosecretory cells are actively synthesizing their material (as suggested by the large nuclei) which is rapidly transported to the corpus cardiacum and allatum to be released into the blood or otherwise utilized. Consequently, the cells contain little material at this time. But in the absence of meat, the release of neurosecretory material from the cells is inhibited, the cells fill up with material, and synthesis is eventually much reduced. Subsequent meat feeding reverses this process (Fig. 7.15). Consequently, it is concluded that the active synthesis and release of the brain factor which affects protease production by the gut is stimulated by meat feeding itself.

Similarly, it can be shown that allatectomy results in the accumulation of material within the neurosecretory cells, whose nuclei are also reduced in size after the operation. The implantation of corpora allata into allatectomized flies causes the neurosecretory cell nuclei to increase in size (Fig. 7.15), as does the implantation of corpora allata into sugar-fed

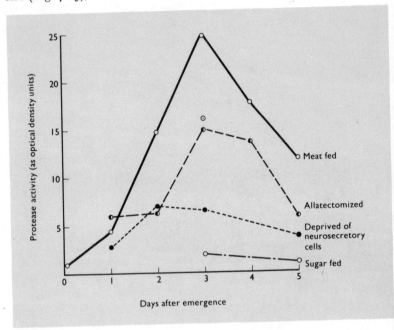

Fig. 7.14 Protease activity in the midgut of *Calliphora erythrocephala*. Normal meat-fed flies show a clear cycle of protease activity which is abolished when the medial neurosecretory cells of the brain are removed, or when the animals are fed only on sugar. After removal of the corpus allatum, the cycle persists but at a lower level. A single observation (⊙) shows that a high protease level can be restored in flies without median neurosecretory cells by the implantation of three pairs of corpora cardiaca-allata. (After Thomsen and Moller[263])

flies. Thus the corpus allatum hormone itself can stimulate the synthetic activity of the neurosecretory cells. Consequently, the effect of allatectomy upon protease production by the gut is most likely indirect, mediated through the activity of the neurosecretory system.

The production of proteases by the cells of the midgut is only one aspect of protein metabolism in these cells. Thus there is a very strong possibility that the control of protease production by the median neuro-secretory cells in *Calliphora* is not just a specific effect upon a particular tissue, but that the neurosecretory cells control the overall protein metabolism of all the tissues in the body. This is to say that the effect of the median neurosecretory cells upon oocyte growth, for example, is *direct* and is not mediated *solely* by its effect upon the availability of nutrients to the oocytes caused by another effect upon protease production by the midgut.

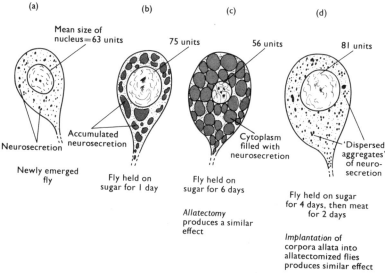

Fig. 7.15 Median neurosecretory cells of the blowfly, *Calliphora erythrocephala* in (a), the newly moulted female, (b) a female fed sugar, but no meat, for 1 day, (c), a fly fed on sugar only for 6 days, or allatectomized, and (d), a fly fed sugar and then meat, or one in which corpora allata are implanted after previous allatec-tomy. Nuclear sizes are given in arbitrary units. The accumulation of inclusions in (b) and (c) indicates that release of neurosecretion is inhibited, and the smaller size of the nucleus in (c) suggests that the synthetic activity of the cell is reduced. Subsequent meat feeding, or corpus allatum implantation (d), raises both the inhibition on release of material and the synthetic activity of the cell. (After Thomsen[264])

In the light of this new evidence about the function of the median neurosecretory cells in the adult, the role of the corpus allatum during egg development in *Calliphora* may now seem rather obscure. The original experiments with the corpus allatum suggested that the gland was involved in oocyte growth. The idea that the corpus allatum hormone may be a metabolic hormone, like that in *Rhodnius*, receives some support

from experiments on the oxygen uptakes of ovariectomized and allatectomized flies[260, 262] (Fig. 7.16). But it must be remembered that these
experiments were carried out before the effects of the corpus allatum
hormone upon neurosecretory cell activity were known. In *Calliphora*,
the reduced oxygen uptake following allatectomy could consequently be
a secondary effect following upon the reduced activity of the neuro-

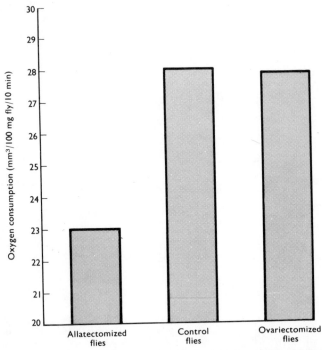

Fig. 7.16 Oxygen consumption in the blowfly, *Calliphora erythrocephala*, after
allatectomy and ovariectomy. Allatectomy causes a significant decrease in oxygen
consumption, whereas ovariectomy has little effect. This suggests that the corpus
allatum hormone has a general metabolic effect in the adult female insect, and is
not merely a gonadotrophic hormone affecting directly the development of the
ovaries. (After Thomsen and Hamburger[262])

secretory system. At the present time, it might be simplest to consider
the corpus allatum hormone as being principally a gonadotrophic hormone, directly influencing the uptake by the oocyte during vitellogenesis
of materials which have been made available by the action of the
neurosecretory hormones.

More recently, evidence has been produced that the size of the corpus

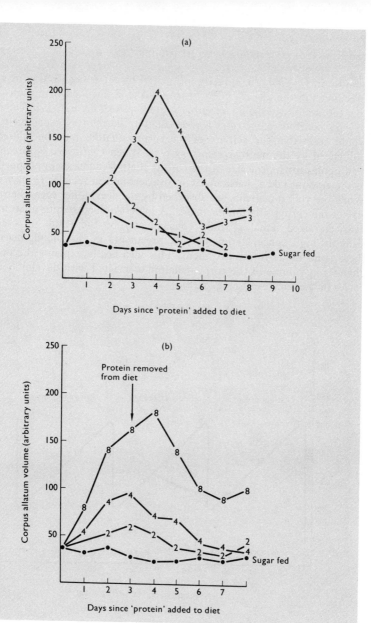

Fig. 7.17 The effects of protein feeding on the size of the corpus allatum in the blowfly, *Calliphora erythrocephala*. (a) Adult female insects are fed the same quantity of protein each day for a different number of days (indicated by the figures which constitute the points on the graph). The longer the time protein remains in the diet, the larger the corpus allatum becomes. In sugar fed flies, the gland remains uniformly small. (b) Flies fed for the same length of time upon different quantities of protein (indicated by the figures constituting the points on the graph). The greater the quantity of protein in the diet, the larger are the corpora allata. (After Strangways-Dixon[254])

allatum in *Calliphora* is directly influenced by the concentration of yolk precursors in the blood[252, 253, 254] (Fig. 7.17). Associated with the claim that the appetite of the fly for either protein or carbohydrate is influenced by the neurosecretory and corpus allatum hormones respectively, it would seem that the activity of the corpus allatum may be *indirectly* influenced by the neurosecretory system. When *Calliphora* is forced to feed upon protein, the size of the corpus allatum increases in proportion to the amount taken, particularly in ovariectomized females (Fig. 7.17). When the yolk precursors are absorbed by the developing eggs, then the size of the corpus allatum decreases. A hypertrophied corpus allatum in ovariectomized females would consequently be the result of a consistently high concentration of yolk precursors in the blood, because there is no tissue to remove them. This idea neatly explains the cyclic changes in volume of the corpus allatum associated with developmental cycles in the oocytes[252] (Fig. 7.18).

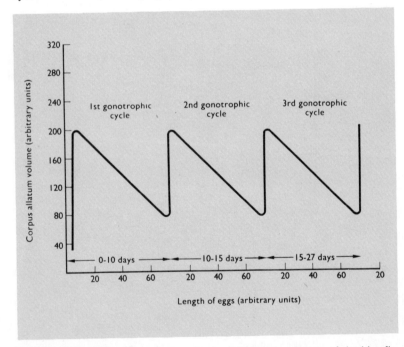

Fig. 7.18 Cyclic variation in the volume of the corpus allatum of the blowfly, *Calliphora erythrocephala*, during the first three gonotrophic cycles. The corpus allatum volume increases during the initial phase of yolk deposition and decreases as the oocytes grow, reaching a minimum value when the eggs are fully developed. (After Strangways-Dixon[252])

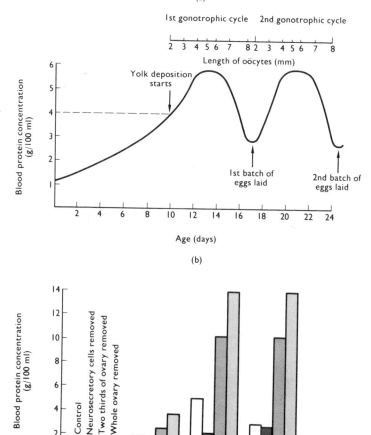

Fig. 7.19 Blood protein concentrations in the adult female desert locust, *Schistocerca gregaria*. (a) Variations during normal egg development. The blood protein concentration increases slowly during the first 8–10 days of adult life, and when it reaches 4 g/100 ml yolk deposition begins in the oocytes. The blood protein concentration continues to increase until the oocytes reach a length of 6 mm, and then decreases to a minimum as the fully developed eggs are ovulated. The pattern is repeated during the subsequent gonotrophic cycles. (b) Variations after cautery of the cerebral neurosecretory cells and after ovariectomy. After the neurosecretory cells are destroyed, the blood protein concentration remains low; after removal of at least two-thirds of the ovaries, the blood proteins increase to a concentration far above normal. (After Hill[144])

In the desert locust, *Schistocerca gregaria*, destruction of the median neurosecretory cells prevents the normal increase in the concentration of blood proteins associated with vitellogenesis[144] (Fig. 7.19). After ovariectomy, the blood protein concentration increases to a very high level, the increase being largely due to two protein groups which are probably vitellogenic (Fig. 7.19 and Plate 2). When radioactively labelled amino acids are injected into the haemolymph, the radioactivity subsequently appears first in the proteins of the fatbody, then in the haemolymph proteins and finally in the yolk proteins of the developing oocytes[145] (Fig. 7.20). This again suggests that the fatbody synthesizes the vitellogenic proteins which are eventually incorporated into the oocytes as yolk. When radioactively labelled amino acids are injected into females whose cerebral neurosecretory cells have been destroyed, the rate of incorporation of the amino acids into protein is considerably reduced (Fig. 7.20). It is concluded that in the locust, as in the blowfly, the

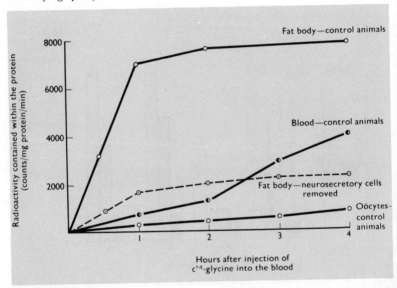

Fig. 7.20   Rates of incorporation of C^{14}-labelled glycine into proteins of the fat body, blood and ovaries of the female desert locust. The rate of incorporation of amino acids into the proteins of the fatbody is used as a measure of the protein synthetic activity of the tissue. The high incorporation rate of control animals suggests a tissue actively synthesizing proteins, while the low rate in animals *minus* their cerebral neurosecretory cells suggests that little synthesis is taking place. Radioactivity does not appear in the blood proteins until after it has appeared in the fatbody proteins, and it finally appears in the ovarian proteins. This sequence suggests that proteins are synthesized in the fatbody, released into the blood, and then taken up from the blood by the ovaries. (After Hill[145])

cerebral neurosecretory cells control oocyte development indirectly by affecting protein synthetic mechanisms in the body.

In the desert locust, removal of the corpora allata once vitellogenesis has begun results in rapid oosorption[142,143] (Fig. 7.21). The effect of

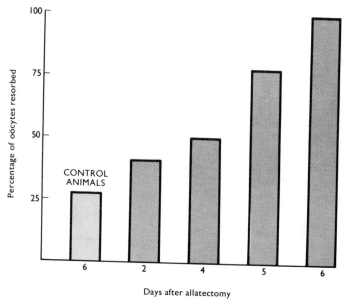

Days after allatectomy

Fig. 7.21   Effects of removing corpora allata in mature female desert locusts, after the initiation of vitellogenesis. Yolk deposition in the oocytes ceases, and that already present is broken down and the oocyte resorbed. Six days after the operation all the developing oocytes in the ovaries have been resorbed. (After Highnam, Lusis and Hill[142])

allatectomy upon the oocytes is obvious within two days; six days after the operation, all the oocytes have been resorbed (Fig. 7.21). This comparatively rapid effect of removal of the corpora allata suggests two things: first, the hormone has a short biological life in the blood, and secondly, because the oocytes respond quickly to its absence, or to its diminished concentration, it is likely that the hormone acts directly upon the oocytes, i.e. it is a gonadotrophic hormone. In some other insects, corpora allata implanted into the ovaries of allatectomized females will induce vitellogenesis only in these oocytes in close proximity to the implanted glands[157]—suggesting strongly a gonadotrophic function of the corpus allatum hormone. But even so, an additional metabolic function for the hormone is also possible: in *Locusta migratoria* it is likely that the corpus allatum hormone determines the *kind* of proteins synthesized by the fatbody. In *Locusta*, the neurosecretory hormone from the brain controls the overall fatbody synthesis, but when the corpus allatum hormone is also present, the fatbody produces specific vitellogenic proteins[208] (Fig. 7.22).

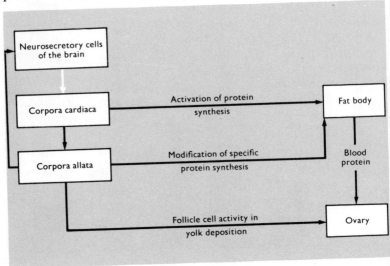

Fig. 7.22 Control of egg development in the migratory locust, *Locusta migratoria*. It has been suggested that in this species a neurosecretory hormone from the brain activates general protein synthesis in the fatbody, but that the corpus allatum hormone directs such synthesis to the production of vitellogenic proteins. The hormone may additionally act as a gonadotrophic hormone to control the uptake of vitellogenic proteins by the oocytes. (After Minks[208])

In some insects, even under the most favourable conditions, some oosorption always occurs. Thus the desert locust, with about 50 ovarioles per ovary, could produce a maximum of 100 eggs in every gonadotrophic cycle. But in practice, about 22% of the oocytes are resorbed some time during their development, the greatest oosorption occurring during the later stages of vitellogenesis (Fig. 7.23). The haemolymph protein concentration begins to fall at this time, and it is possible that oosorption results from competition between the developing oocytes for available protein. When female desert locusts are *partially* ovariectomized, the remaining oocytes develop in conditions of increased protein 'availability'—in fact the haemolymph protein concentration can be very considerably increased (Fig. 7.19). Under these conditions, the percentage oosorption is reduced by about half, compared with normal

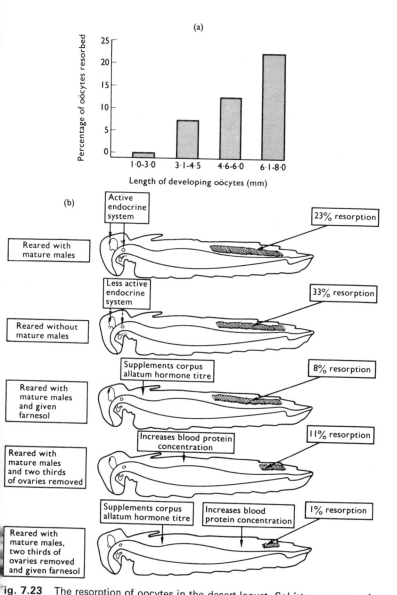

Fig. 7.23 The resorption of oocytes in the desert locust, *Schistocerca gregaria*. (a) A proportion of the oocytes are resorbed even during normal oocyte development: 8% during the early stages and a further 15% when the oocytes are between ·5 and 8·0 mm in length. (b) If *either* vitellogenic proteins *or* corpus allatum ormone are made available in larger quantity, oocyte resorption is decreased; *both* are increased oocyte resorption is virtually eliminated. It is concluded at oocyte resorption in the desert locust is due to competition by the developing ocytes for vitellogenic proteins and corpus allatum hormone.

animals (Fig. 7.23). But even with three-quarters of the ovaries removed, about 10% of the remaining oocytes are resorbed. Therefore, competition between oocytes for protein can account for only some of the oosorption which occurs in normal animals. But when such partially ovariectomized females are treated with corpus allatum hormone mimics (see p. 225), then oosorption is completely eliminated (Fig. 7.23), and in fact the penultimate and even the antepenultimate oocytes in the ovarioles lay down yolk (Fig. 7.24). In unoperated females, the percentage oosorption

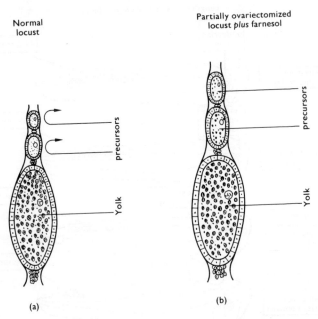

Fig. 7.24 Yolk deposition in the oocyte of the desert locust, *Schistocerca gregaria*. In the normal locust (a) yolk is deposited only in the terminal oocyte (that nearest the oviduct) in each ovariole. But in a female with two-thirds of her ovaries removed, with the juvenile hormone mimic farnesol added (b), not only the terminal oocytes but the penultimate and ante-penultimate oocytes also deposit yolk. In such females, the haemolymph protein concentration and juvenile hormone (or its mimic) concentration are both high and presumed competition for these materials by the developing oocytes is consequently greatly reduced. Compare Fig. 7.23.

is reduced—but not eliminated—by treatment with such corpus allatum hormone mimics. It is therefore concluded that in the desert locust, the oosorption which occurs normally is the result of oocyte competition for *both* protein and corpus allatum hormone.[143] Since the materials from resorbed oocyte are presumably made available to the developing oocytes, this system has the obvious advantage of enabling a proportion of the oocytes to develop completely when conditions are not entirely favourable.

In insects such as the blowfly, locusts and the cockroach, cerebral neurosecretion together with the corpus allatum hormone is necessary for oocyte development. In *Rhodnius* and many Lepidoptera, neurosecretory hormones play no part in vitellogenesis. In the milkweed bug, *Oncopeltus fasciatus*, and some stick insects (Phasmidae), removal of the cerebral neurosecretory cells has no immediate effect upon oocyte development but the total fecundity of the operated females—the number of eggs they produce during their reproductive lives—is considerably reduced (Fig. 7.25). At present, the endocrine control of oocyte develop-

	Longevity (days)	Egg yield (total number of eggs produced)
Control animals	$88 \cdot 3 \pm 14 \cdot 1$	$1263 \cdot 4 \pm 252 \cdot 4$
Animals without cerebral neuro-secretory cells	$78 \cdot 3 \pm 12 \cdot 0$	$573 \cdot 5 \pm 121 \cdot 2$

Fig. 7.25 Neurosecretion and reproduction in the milkweed bug, *Oncopeltus fasciatus*. Removal of the cerebral neurosecretory cells does not prevent egg development in this species, but the fecundity of the female is considerably reduced.

ment has been examined in too few insects for any general principles to be established.[134] The kinds of endocrine control of oocyte development illustrated in Fig. 7.26 may have to be considerably modified in the future.

In those insects in which both cerebral neurosecretion and corpus allatum hormone combine to control the progress of oocyte development, the hormonal relationship is complicated by an interaction between the two endocrine centres. In the blowfly, the activity of the cerebral neurosecretory cells, as measured by their nuclear volumes, is enhanced by the presence of the corpus allatum hormone (p. 122, Fig. 7.15). In the desert locust, similarly, release of neurosecretion is retarded after

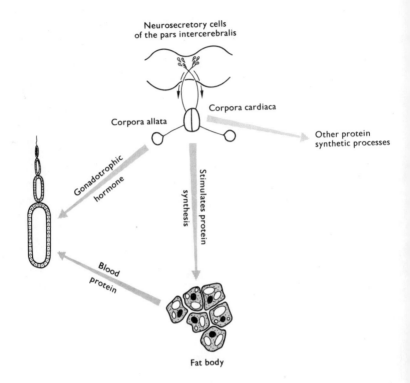

Fig. 7.26 A general scheme for the endocrine control of egg development in an insect such as the desert locust.

allatectomy, and accelerated when the glands are reimplanted. Moreover, in *Locusta migratoria* the activity of the corpora allata is said to be inhibited by the secretion from the A- and B-cells in the brain, and stimulated by the secretion from the C-cells[113] (Fig. 7.27). Such interactions between the cerebral neurosecretory system and the corpora allata obviously make it difficult to interpret the effects of removal and reimplantation of one or the other endocrine centre. Indeed, it is possible that some of the metabolic effects ascribed to the corpus allatum hormone may be the results of alterations in neurosecretory activity consequent upon allatectomy and reimplantation of the corpora allata. This rather confusing situation will not be completely resolved until pure preparations of both hormones are available for experimental investigations.

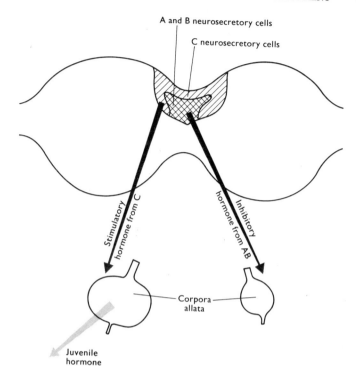

A and B neurosecretory cells

C neurosecretory cells

Stimulatory hormone from C

Inhibitory hormone from AB

Corpora allata

Juvenile hormone

Fig. 7.27 Possible control of the corpora allata by neurosecretory factors in the migratory locust, *Locusta migratoria*. The C neurosecretory cells in the brain stimulate the corpora allata to produce juvenile hormone, the A and B cells inhibit the glands. *Isolated* corpora allata can be stimulated or inhibited, so the neurosecretory factors do not necessarily pass along the nerves to the glands, but can be released into the haemolymph. These conclusions have been reached by selective destruction of the A, B and C cells: their arrangement in *Locusta* makes this possible. (After Girardie[112])

ENDOCRINE CONTROL OF ACCESSORY GLAND DEVELOPMENT

In many female insects, the accessory glands are small and thin in the newly emerged adult, increasing in diameter and length as the oocytes develop and producing secretions which are poured onto the eggs during ovulation and oviposition. After allatectomy, the accessory glands remain small, only growing if corpora allata are subsequently reimplanted (Fig. 7.28).

But it must not immediately be assumed that the corpus allatum hormone directly affects accessory gland development. Growth of the accessory glands is associated with the development of the oocytes, and these do not develop after allatectomy. Consequently, the retarded accessory gland development after allatectomy *could* be due to the absence of a necessary factor from the undeveloped oocytes.

The obvious way to test whether the corpora allata control accessory gland development directly, or indirectly through the ovaries, is to ovariectomize the newly emerged females. In such females, the accessory glands develop more or less normally. Consequently, it is concluded that

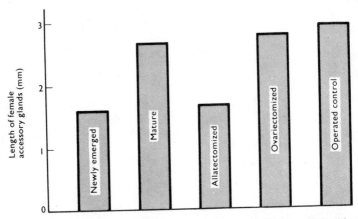

Fig. 7.28 Comparison of the lengths of the accessory glands of the blowfly, *Calliphora erythrocephala*, after allatectomy and ovariectomy. Allatectomy prevents the normal increase in length of the glands which accompanies maturation (see Fig. 7.7), but ovariectomy does not. Consequently, the ovaries themselves do not influence accessory gland size. The glands and ovarian development are controlled independently by the corpus allatum.

in insects the development of the accessory glands and the oocytes is correlated by the corpus allatum hormone, and not by the accessory glands being dependent upon ovarian secretions for their growth. This contrasts markedly with the situation in the vertebrates, where glandular and other changes in the reproductive tract are controlled by ovarian hormones. In fact, there is no indication that the insect ovary is a source of any hormone.

In some insects, the female accessory glands respond to a lower concentration of corpus allatum hormone than do the oocytes. Consequently as the corpora allata become active after adult emergence, the accessory glands are developmentally slightly in advance of the oocytes, ensuring that their secretions are present in quantity when the oocytes are ovulated.

CYCLIC DEVELOPMENT OF OOCYTES

In many insects, the terminal oocytes in all the ovarioles develop synchronously and all the eggs are laid at one time; after ovulation from the ovarioles, their places are taken by the previously penultimate oocytes, which continue development and are laid. The originally ante-penultimate oocytes then become terminal and begin their development, and so on. Consequently, batches of eggs are laid at more or less regular intervals, and each laying represents the end of a particular gonotrophic cycle. After ovulation, the follicle cells which surrounded each oocyte contract to form a ring of tissue, often pigmented, around the base of each ovariole.[198] This ring of pigmented tissue is sometimes called the *corpus luteum*, although it must be clearly understood that it has no functional analogy whatsoever to the corpora lutea of vertebrates. The number of corpora lutea at the base of each ovariole can often be used to determine how many gonotrophic cycles have occurred (Fig. 7.29).

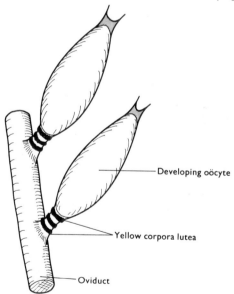

Fig. 7.29 Diagram of the corpora lutea at the bases of two ovarioles in the desert locust, *Schistocerca gregaria*. Three gonotrophic cycles have been completed, and the developing oocytes therefore represent the fourth cycle.

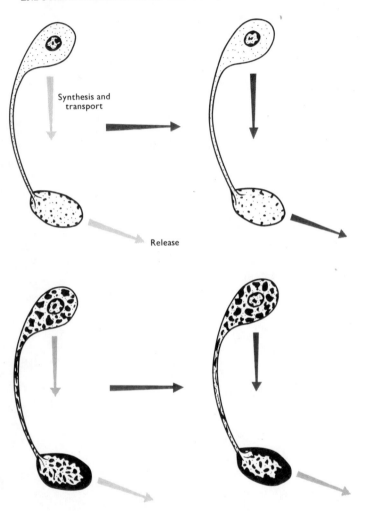

Fig. 7.30 Histological differences between insect neurosecretory systems (diagrammatic). The density of shading of the broad arrows represents degrees of synthesis and transport, and release of neurosecretion.
Upper row : the neurosecretory systems contain relatively little histologically demonstrable material, but synthesis and release of neurosecretion can increase considerably without altering the appearance of the system so long as they remain in balance.
Lower row : the systems are full of material, but again synthesis and release of neurosecretion can vary widely without altering the histological appearance so long as synthesis exceeds release. (After Highnam[136])

The volumes of the corpora allata, the size of their nuclei, and the volume of cytoplasm per nucleus fluctuate in relation to each gonotrophic cycle[252] (Fig. 7.18). These fluctuations are said to reflect varying secretory activity in the glands, the largest glands being most active. However, in some insects under certain conditions, it has recently been shown that the relationship between size and activity does not always hold true. A large corpus allatum, for example, could be *storing* hormone and not releasing it into the blood. Now that the corpus allatum hormone has been identified, it should be possible in the future to measure its concentration in the blood directly.

Variations in the activity of the cerebral neurosecretory system are also associated with each gonotrophic cycle. The criteria used to determine activity are again largely histological, depending upon the amounts of stainable material present in the neurosecretory system at different times. Neurosecretions are synthesized by the cells in the brain, transported along the neurosecretory cell axons and stored for a longer or shorter period of time in the corpora cardiaca. Clearly, the content of neurosecretion within the system at any time will depend upon its relative rates of synthesis, transport and release, which can all vary independently. Histological pictures from material fixed at an instant in time are consequently of little use in estimating variations in the essentially dynamic process of turnover of neurosecretion[133, 136] (Fig. 7.30).

Neurosecretory proteins are rich in the sulphur containing amino acids cystine and cysteine (Fig. 7.31). Consequently, radioactively-

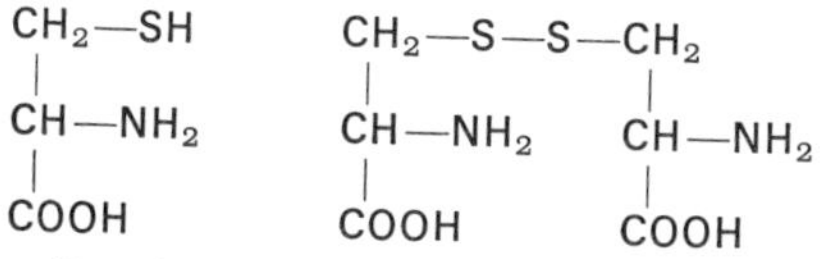

$$CH_2-SH \qquad CH_2-S-S-CH_2$$
$$CH-NH_2 \qquad CH-NH_2 \qquad CH-NH_2$$
$$COOH \qquad COOH \qquad COOH$$
$$\text{Cysteine} \qquad \qquad \text{Cystine}$$

Fig. 7.31 Formulae for cystine and cysteine. When the sulphur atoms are radioactive, they label protein molecules into which the amino acids are incorporated.

labelled S-amino acids injected into the blood, are incorporated preferentially into neurosecretory proteins. The rate of incorporation of the amino acids by the cerebral neurosecretory cells, and their rate of release from the corpora cardiaca, can be measured by auto-radiographic and other means, and the *relative* activity of the neurosecretory system can then be estimated at different times during the gonotrophic cycle.[133, 137]

What causes these cyclic changes in the cerebral neurosecretory system and the corpora allata? A large number of factors influence the activity of the cerebral neurosecretory system (see Chapter 14) amongst which feeding and mating have marked stimulatory effects (Plate 7b). The first gonotrophic cycle in the desert locust could therefore be initiated and maintained by the voracious feeding in which the females indulge after emergence, together with copulation and other factors (p. 244) which combine to stimulate the production and release of neurosecretory hormones. Because of their relationship with the cerebral neurosecretory system, activation of the corpora allata would follow. In the blowfly, meat feeding is necessary for the release of material from the cerebral neurosecretory cells (p. 121; Fig. 7.15). In *Rhodnius*, a blood meal activates the corpus allatum, although whether this is direct or mediated through the neurosecretory system is not known.

But what causes the decline in activity of the parts of the endocrine system towards the end of a gonotrophic cycle? At present, the answer to this question is unknown. In some insects, ovariectomy causes hypertrophy of the corpora allata, either because the glands now store their hormone or because they are hyperactive—striving to achieve a result which is no longer possible. Ovariectomy can also inhibit the release of neurosecretion from the corpora cardiaca (Plate 7). It is possible that the increasing concentrations of haemolymph metabolites which follow ovariectomy in some way act back on the source of the metabolic hormone which is responsible for their production. A similar situation arises in normal animals when the oocytes have completed vitellogenesis and no longer abstract metabolites from the haemolymph at their previous rate. It is also possible that in some insects the *removal* of metabolites from the blood will itself affect endocrine activity. But it is not inconceivable that nervous impulses from an abdomen stretched by its content of large oocytes could affect the endocrine system. In the viviparous cockroach, *Leucophaea maderae*, the presence of developing young in the 'uterus' stimulates sensory receptors whose signals eventually inhibit the activity of the corpora allata.[85, 86]

By analogy with the situation in the vertebrates, the production of a hormone by the developing oocytes which reduces the activity of the cerebral endocrine system might seem the obvious way to achieve the known result. Indeed, the existence of such a hormone has been claimed in some insects.[212] But the evidence is not good, and it is likely that in these insects, too, a rather more complicated feed-back mechanism will eventually be found.

In the desert locust, the act of egg-laying itself stimulates the cerebral neurosecretory system once more into activity, the maintenance of this activity subsequently depending upon continued feeding and perhaps other factors. Thus the terminal process of one gonotrophic cycle triggers off the beginning of the next. It is likely that neurosecretory hormones released during oviposition facilitate the oviducal and other movements necessary for the movement of the eggs out of the body. In *Rhodnius*, neurosecretory factors are also necessary for egg-laying[75]; the process has not been examined in detail in any other insects.

ENDOCRINE CONTROL OF REPRODUCTION IN THE MALE INSECT

The endocrine control of oocyte development applies strictly to vitellogenesis—oogenesis within the germarium is not affected by hormones. It is not surprising, therefore, that in the adult male insect spermatogenesis proceeds independently of any hormonal influence. However, differentiation of the testes, including the development of spermatids from spermatogonia, proceeds rapidly during the last moult. This process is part of the general metamorphic transformation of the body, and it is likely, therefore, that it is controlled by the same combination of hormones that effects metamorphosis (see p. 81).

Accessory glands in the male insect secrete protein and other materials which form the spermatophore, in those species where it occurs, and also the bulk of the seminal plasma which carries the sperm. In many insects growth and development of the male accessory glands is controlled by the corpus allatum hormone in an exactly similar manner to that of the female accessory glands (p. 135*ff*).

8

Endocrine Mechanisms in the Insecta—III

Other Endocrine-controlled Processes

HORMONES AND DIAPAUSE

Many insects avoid unfavourable periods by entering diapause, a state of arrested development. The most familiar examples are the over-wintering chrysalids, or pupae, of many British butterflies and moths, but diapause may also occur in the egg, in the larval stages and also in the adult. Usually only one diapause intervenes during the life of an individual, but species with more than one diapause in their life-cycles are not unknown.

Diapause is characterized by a considerably diminished metabolism, by the presence of large reserves of fat, sometimes associated with pro-tein and carbohydrate, by a marked resistance to desiccation (important particularly for aestivating insects in arid tropical regions) and by ex-treme cold-resistance, reflected in an ability to withstand low tempera-tures (from $0°$ to $-10°$) that are usually lethal to an insect, and even temperatures as low as $-40°$. Diapause must therefore be anticipated by the building-up of special reserves, and the formation of protective devices such as wax layers to minimize evaporation. Consequently, it is easily distinguished from the torpor induced by low or high temperatures or desiccation, which cause pathological changes resulting in death.

In insects with a long period of post-embryonic growth, lasting a year or more, diapause may intervene at the correct time regardless of environmental conditions. Such diapause is termed **obligatory** and is normally associated with univoltine species, i.e. those having one generation a year. In polyvoltine species, which have two or more generations a year, diapause occurs only in that generation which spans the unfavourable period. Thus the cabbage white butterfly, *Pieris brassicae*, lays eggs in early summer, larvae emerge from these and develop directly into adults through several instars and through a pupal stage. But the eggs from these adults hatch into larvae which develop later in the summer and produce a diapausing pupa which overwinters. Such a diapause is called *facultative*, since it depends for its appearance upon the appropriate environmental conditions.

Many hypotheses have been proposed to account for the intervention of diapause in insect life-histories, the most popular at one time being self-poisoning or auto-intoxication of the insect by a gradual accumulation of metabolic waste materials. But it is now known that post-embryonic diapause results from the temporary cessation of activity of those parts of the endocrine system which control growth and development (Chapter 6).

In the pupa of the giant silkmoth, *Hyalophora cecropia*, diapause can be terminated by a prolonged period of chilling, for example 6 weeks at 3°C, followed by 2 weeks at a higher temperature such as 25°C.[286] The period of adverse conditions is the time required for specific physiological changes to occur, which make possible the recommencement of normal development. It is therefore called the period of diapause development and is characteristic of most insect diapauses.[2]

If the brain of *Hyalophora* is removed at any time during pupal diapause development, then the decerebrate individual will not terminate diapause. But if the brain from an individual chilled for 6 weeks is implanted into an unchilled diapausing pupa, then development is eventually resumed in the normal way. Thus the effect of chilling is solely upon the brain, which develops a competence to terminate diapause.

If a chilled brain is implanted into the isolated pupal abdomen of *Hyalophora*, diapause is not terminated (Fig. 8.1). But if the chilled brain is implanted into a pupa containing the thoracic glands, or if the glands are implanted together with the chilled brain then normal development will ensue[287] (Fig. 8.1). Clearly, diapause is due to failure of the cerebral neurosecretory cells to produce thoracotrophic hormone after pupation in *Hyalophora*, the absence of ecdysone preventing further development. Convincing evidence for this theory is that implanted activated thoracic glands will terminate diapause in pupae of *Hyalophora*,

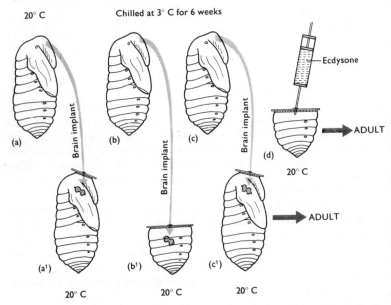

Fig. 8.1   Termination of pupal diapause in *Hyalophora cecropia*. The brain from an unchilled pupa (a) implanted into an unchilled pupa (a′) does not end diapause and initiate development. A brain from a chilled pupa (c) will terminate diapause when implanted into an unchilled individual (c′). But a chilled brain (b) will not terminate diapause when implanted into an unchilled isolated abdomen (b′), but ecdysone injected into such an abdomen (d) will terminate diapause and produce an adult abdomen. These results suggest that the effect of chilling is upon the brain, which subsequently releases thoracotrophic hormone to stimulate the thoracic glands to produce ecdysone. (After Williams[287])

as will injections of pure ecdysone. During diapause development in *Hyalophora*, therefore, the cerebral neurosecretory cells initiate the production of thoracotrophic hormone which exerts its effect when higher temperatures prevail.

It is likely that the majority of both pupal and larval diapauses result from the failure of the cerebral neurosecretory cells to produce thoracotrophic hormone. But adult diapause, characterized by lack of development of the gonads, and particularly the ovaries, often occurs without any associated quiescence of the parents and can be due to other factors. Thus the water beetles *Dytiscus marginalis* and *Dytiscus semisulcatus* enter reproductive diapause in March and October respectively. During diapause, the oocytes with their accompanying nurse cells surrounded by follicle cells, continually pass down the ovarioles. But when the oocyte is about 1–2 mm in diameter, it is resorbed by the follicle cells which then form a corpus luteum. The same thing happens to the oocyte next in line in each ovariole.[157]

Oosorption during diapause in these water beetles occurs at the time the oocytes lose connection with their nurse cells. This, of course, mimics exactly the effects of allatectomy upon other insects with meroistic ovarioles (p. 113), and it might be suspected that reproductive diapause in *Dytiscus* was the result of failure of the corpora allata to secrete its gonadotrophic hormone. In fact, the implantation of 10 pairs of active corpora allata into the diapausing beetles induces normal egg development and oviposition, and allatectomy of non-diapausing *Dytiscus* results in oosorption rather similar to that which occurs during diapause, except that eventually the ovarioles themselves regress. Reproductive diapause in *Dytiscus* is thus due to the relative inactivity of the corpora allata,[157] although because the changes in the ovaries during diapause are not as pronounced as those following allatectomy, the glands cannot be *completely* inactive.

During reproductive diapause in the Colorado beetle, *Leptinotarsa decemlineata*, the ovaries show changes which mimic exactly the effects of allatectomy, and again the inactivity of the corpora allata must be of importance in controlling diapause.[285] But in this species, the implantation of corpora allata into the diapausing females will only induce full oocyte development for a short time after the induction of diapause. If the environmental conditions which bring about diapause are continued, brain implants from active individuals are necessary to break diapause. In the Colorado beetle, therefore, the onset of diapause is brought about by a rapid inhibition of the corpora allata, but the diapause is strengthened and maintained by the subsequent inactivity of the cerebral neurosecretory cells. In the bug *Pyrrhocoris apterus*, reproductive diapause is similarly maintained by the absence of circulating neurosecretory and corpus allatum hormones.[247]

During the reproductive diapause in the Egyptian grasshopper, *Anacridium aegyptium*, which has panoistic ovarioles, there is no turnover of oocytes—development to a certain stage followed by oosorption—like that found in species with meroistic ovarioles (see above). Instead, the oocytes remain small and undeveloped and never show signs of vitellogenesis. This condition again recalls the different effects of allatectomy in insects with panoistic or meroistic ovarioles, and in *Anacridium* inactivation of the corpora allata is certainly one aspect of reproductive diapause. But the very small oocytes of *Anacridium* during diapause resemble more closely the oocytes of non-diapausing females which have had their cerebral neurosecretory cells removed, than those of allatectomized non-diapausing females in which the oocytes are much larger although still lacking yolk.[104] In *Anacridium*, therefore, failure of the cerebral neurosecretory cells is likely to be the primary cause of reproductive diapause, inactivity of the corpora allata being a secondary consequence.

Post-embryonic diapause in insects is thus due to the *absence* of developmental hormones. But egg diapause, on the contrary, is induced by the *presence* of an inhibitory hormone, secreted by the female and affecting her developing oocytes.

Egg diapause is facultative in many races of the silkmoth, *Bombyx mori*. According to how they are reared (see below) adults will lay either diapause or non-diapause eggs. In *Bombyx*, oocyte development begins in the pupal stage, and this is when the diapause hormone acts.[97] The hormone could function by increasing the permeability of the eggs to substances which subsequently inhibit development.

When the brain is removed from diapause-producing pupae 60 hours after pupation, only about 17% of the eggs eventually produced enter diapause. This proportion increases progressively the longer brain removal is delayed after pupation (Fig. 8.2). Removal of the brain shortly after pupation in non-diapause producing pupae causes about 57% of the eggs eventually laid to *enter* diapause. This proportion diminishes progressively the longer brain removal is delayed (Fig. 8.3).

Removal of the sub-oesophageal ganglion from diapause producing pupae shortly after pupation completely eliminates diapause from the eggs eventually produced; when the operation is delayed, a progressively larger proportion of eggs enter diapause (Fig. 8.4). The same operation on non-diapause producing pupae has no apparent effect; all the eggs are without diapause just as they would have been if the ganglion had not been removed (Fig. 8.5).

How can these results be explained? The effects of brain removal from a diapause producing pupa might suggest that the brain is the source of the diapause hormone, but this cannot be so because the same operation on non-diapause producing pupae causes some of the eggs to enter diapause. When brains from either diapause producing or non-diapause producing pupae are transplanted into either kind of pupa, they are absolutely without effect. But the lack of diapause eggs after removing the sub-oesophageal ganglion suggests that this is the source of the diapause hormone in diapause producing pupae, and that it is absent from non-diapause producing pupae. This is confirmed by transplanting sub-oesophageal ganglia from diapause producing to non-diapause producing pupae: a large proportion of the eggs laid enter diapause (Fig. 8.6). In fact, the diapause hormone is secreted by a single pair of large neurosecretory cells in the sub-oesophageal ganglion.[97]

But the brain is clearly involved also in diapause induction. A clue to its function is given by the experiment described above: when

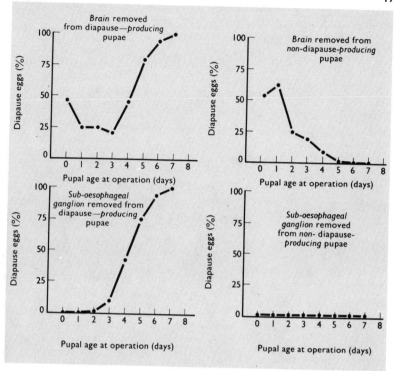

Fig. 8.2 *top left*   Effects upon egg diapause of removing the brain at different times in pupal life of a diapause producing generation of *Bombyx mori*. In general, the longer the brain remains in place, the greater the proportion of diapause eggs produced. This result might suggest that the brain produces a diapause factor—but see Fig. 8.3.

Fig. 8.3 *top right*   As for Fig. 8.2, but in a non-diapause producing generation of *Bombyx mori*. Although the pupa normally would not produce diapause eggs, brain removal in early life *induces* the formation of a proportion of diapause eggs. Contrary to the suggestion in Fig. 8.2, the brain cannot be producing the diapause factor, which must come from elsewhere.

Fig. 8.4 *bottom left*   As for Fig. 8.2, but the sub-oesophageal ganglion is removed. Again, the longer the ganglion remains in place, the larger the proportion of diapause eggs eventually produced. So the sub-oesophageal ganglion could be the source of a diapause factor.

Fig. 8.5 *bottom right*   As for Fig. 8.4, but in a non-diapause producing generation of *Bombyx mori*. These results are consistent with those of Fig. 8.4: if no diapause factor is produced by the sub-oesophageal ganglion in the non-diapause producing generation, then removal of the ganglion would have no effect upon the induction of egg diapause. So it may be concluded that the sub-oesophageal ganglion is the source of the diapause factor, although Figs. 8.2 and 8.3 suggest that the brain is also involved. (Figs. 8.2–8.5 based on data from Fukuda[97])

sub-oesophageal ganglion from a diapause producing pupa is transplanted to a non-diapause producing individual, not *all* the eggs produced enter diapause. Moreover, when a sub-oesophageal ganglion from a non-diapause producing pupa is transplanted to a similar non-diapause producer, then a proportion of eggs *enter* diapause as though the ganglion came from a diapause producer (Fig. 8.6). The function of the brain is now clear: in diapause producers the brain stimulates the production

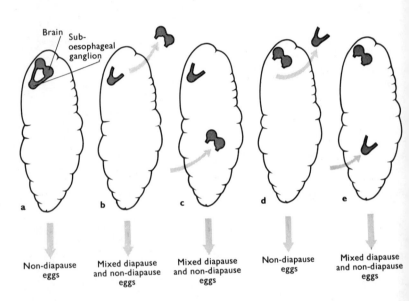

Fig. 8.6(a) The effects of brain and sub-oesophageal ganglion removal and implantation in pupae of a non-diapause producing generation of *Bombyx mori*. (a) Control pupae, brain and ganglion *in situ*: no eggs eventually produced have diapause. (b) Brain removed: although no diapause eggs are normally produced, when the brain is removed early in pupal life, mixed diapause and non-diapause eggs are produced (compare Fig. 8.3). (c) When the brain is reimplanted to another site, mixed diapause and non-diapause eggs are still formed. The connections between the brain and the ganglion must be intact, as in a, for all non-diapause eggs to be formed. (d) Sub-oesophageal ganglion removed: all non-diapause eggs produced. (e) Sub-oesophageal ganglion reimplanted to a different site: mixed diapause and non-diapause eggs produced as in c.

Thus a proportion of diapause eggs are formed in the non-diapause pupae only when the sub-oesophageal ganglion is present either without the brain, or when the connectives with the brain are severed. This suggests that the source of the diapause factor is the sub-oesophageal ganglion, but that normally its release is prevented by the brain *so long as the nervous connections between the two are intact*.

and/or release of hormone from the sub-oesophageal ganglion; in non-diapause producers the brain inhibits hormone production and/or release from the sub-oesophageal ganglion[97] (Fig. 8.7). Merely cutting the cir-

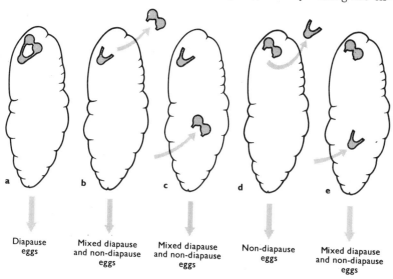

| Diapause eggs | Mixed diapause and non-diapause eggs | Mixed diapause and non-diapause eggs | Non-diapause eggs | Mixed diapause and non-diapause eggs |
| a | b | c | d | e |

Fig. 8.6(b) As for Fig. 8.6(a), but in a diapause producing generation. (a) Brain and sub-oesophageal ganglion intact (control): all diapause eggs produced. (b) Brain removed: a proportion of non-diapause eggs produced (compare Fig. 8.2). (c) Brain reimplanted to a different site: a proportion of non-diapause eggs produced. (d) Sub-oesophageal ganglion removed: non-diapause eggs produced. (e) Ganglion reimplanted to a different site: a proportion of non-diapause eggs produced.

Thus again it may be concluded that the sub-oesophageal ganglion is the source of a diapause factor, but that it produces its maximal effect *only when the connectives between ganglion and brain are intact*. In effect, the brain stimulates the production of a diapause factor by the sub-oesophageal ganglion. (After Fukuda[97])

cumoesophageal connectives between the brain and sub-oesophageal ganglion will produce mixed diapause and non-diapause eggs in both diapause producing and non-diapause producing pupae, because stimulation of hormone production is interrupted in the former, and inhibition removed in the latter. The brain will still exert its stimulatory or inhibitory effect upon the sub-oesophageal ganglion when transplanted into other individuals, so long as the connectives remain intact. A brain and a sub-oesophageal ganglion implanted separately act as if the ganglion alone were implanted, producing mixed diapause and non-diapause eggs.

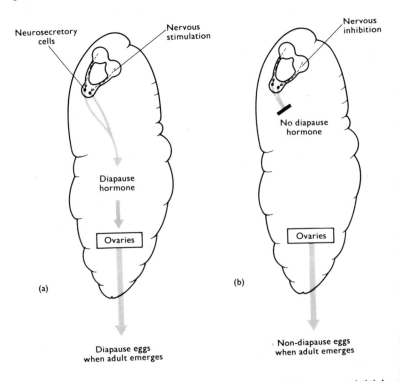

Fig. 8.7 Summary of diapause induction in the eggs of *Bombyx mori*. (a) In a diapause producing pupa, the brain stimulates the production of diapause hormone by neurosecretory cells in the sub-oesophageal ganglion. The hormone is absorbed by the developing eggs, eventually to cause diapause when the eggs are laid. (b) In a non-diapause producing pupa, the brain inhibits the sub-oesophageal ganglion neurosecretory cells from producing diapause hormone, and in its absence the eggs develop without diapause. (After Fukuda[97])

The nature of the control exercised by the brain upon the neuro-secretory cells of the sub-oesophageal ganglion is unknown. It is presumably nervous, from some endogenous centre, because the brain can act in the absence of any afferent input—as for example, when it is transplanted—since all its nerves are necessarily severed during such an operation. But an endocrine control, perhaps by neurosecretory axons travelling directly to the sub-oesophageal ganglion neurosecretory cells from some centre in the brain, cannot be entirely ruled out, although no trace of such axons has yet been found.

What determines whether *Bombyx* pupae will be diapause producers or non-diapause producers? The most important factor turns out to be the temperature at which the eggs are incubated. When the embryos develop at a temperature of about 15°C, then the emergent larvae and their pupae are non-diapause producers—the adult laying eggs without diapause. But incubating the eggs at 25°C gives diapause producing larvae and pupae—the adults will lay all diapause eggs. Incubation at 20°C produces three kinds of larvae and pupae—exclusively diapause producers, exclusively non-diapause producers, and also individuals which will eventually produce mixed diapause and non-diapause eggs. The proportions of these three types vary with the conditions of larval rearing. But most important, long periods of light during egg incubation at 20°C increase the proportion of diapause producers; their absence increases the proportion of non-diapause producers.

The significance of this mechanism for a bivoltine species in nature is obvious (Fig. 8.8). Embryonic development in the low temperatures and short photoperiods of early spring produces eventually adults which lay

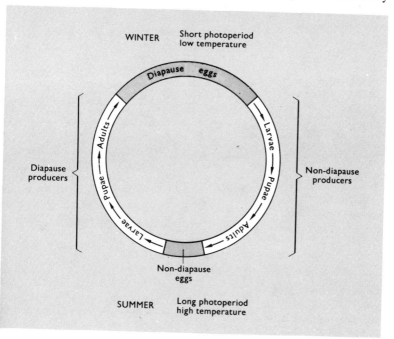

Fig. 8.8 Life cycle of a bivoltine race of *Bombyx mori* with egg diapause. Environmental conditions for the embryos developing in the summer induce diapause in the winter eggs; those for winter embryos prevent diapause in the summer eggs.

non-diapause eggs. These develop directly into the second generation, which takes advantage of the favourable conditions of summer. But the embryos in the non-diapause eggs develop in the high temperatures and long photoperiods of summer—and therefore the adults eventually produced will lay diapause eggs to avoid the unfavourable winter. What is less obvious, of course, is how the incubation temperature of the embryo affects an endogenous nervous centre in the brain which then either stimulates or inhibits the production of a neurosecretory hormone in the pupa.

Although *Bombyx* is rather unusual in that the conditions which determine diapause act upon a stage almost an entire generation before development is arrested, yet it still illustrates an important principle: that when an unfavourable period is avoided by diapause, prior preparation has to be made, and consequently diapause is determined by condi-

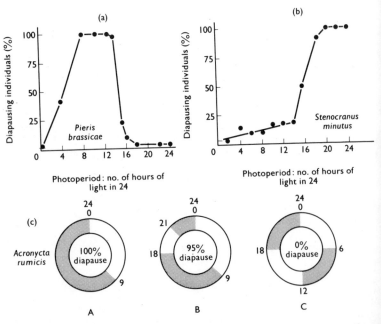

Fig. 8.9 (a) A long day insect, *Pieris brassicae*, in which short day conditions induce diapause. (b) A short day insect, *Stenocranis minutus*, in which long days induce diapause. (c) In the noctuid, *Acronycta rumicis*, pupal diapause is induced by short photoperiods (A), but the dark phase must last for at least 9 *continuous* hours, or not all individuals enter diapause (B). When a 12 hour dark period is interrupted by 6 hours light, diapause is prevented (C). (a and b after Danilevskii[67]; c after Lees[187])

tions which invariably precede the unfavourable period and not by the period itself.

Diapause can be triggered by a number of environmental factors, including temperature and the quality or quantity of food, which vary seasonally.[187] But these factors are intrinsically variable, and are not very good indicators of season. But one environmental factor varies predictably during the course of the year: this factor is photoperiod, the number of hours of light in 24 hours. It is not surprising that many insects react to particular photoperiods which determine the onset of diapause (Fig. 8.9). Even so, the action of photoperiod is often closely linked with temperature, a particular day length inducing diapause only when the temperature lies between rather close limits[67] (Fig. 8.10).

In many insects, diapause results from exposure of the sensitive stages to *short* photoperiods (Fig. 8.9). *Bombyx* is exceptional in that long photoperiods induce diapause. High temperatures usually avert diapause, while low temperatures induce it (Fig. 8.10), although *Bombyx* (see above) is again exceptional in this respect.

The stage in the life-history which is sensitive to these factors which induce diapause is usually fixed and characteristic for each species (Fig. 8.11). In *Bombyx*, the egg is most sensitive to photoperiod, although the early larval instars also show some sensitivity. In the giant silkmoths

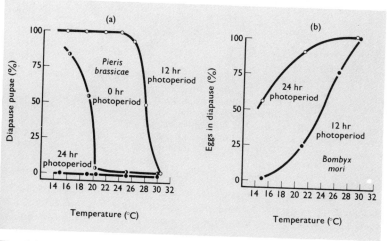

Fig. 8.10 Effects of temperature and photoperiod upon diapause. (a) In the pupa of *Pieris brassicae* when reared at different temperatures and three light regimes. Increasing temperature opposes short photoperiods in inducing diapause. (b) Egg diapause in *Bombyx mori* reared at different temperatures under two light regimes. Increasing temperature *supplements* the longer photoperiod in inducing diapause.

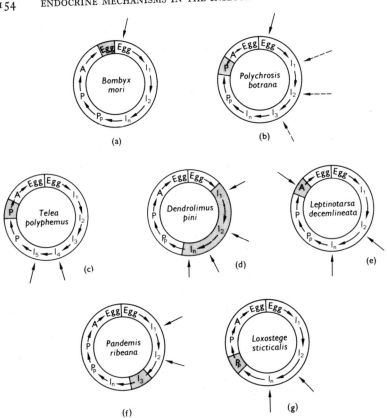

Fig. 8.11 Sensitive stages (shown by arrows) in the life histories of various insects during which environmental stimuli (particularly photoperiod) induce diapause (stippled). In *Bombyx mori* (a) almost a whole generation intervenes between the sensitive stage and the diapause stage. Usually the diapause stage follows the sensitive stage(s) (b, c, e, f, g) although they may be the same (d).

Telea polyphemus and *Philosamia cynthia*, only the 4th and 5th instars react to the short photoperiods which induce pupal diapause. In *Antheraea pernyi*, all the larval stages are sensitive to some extent, but the later ones are more so and the effects are cumulative. In the tomato moth *Diataraxia oleracea*, the most sensitive period is no more than 2 days at the end of the penultimate larval instar. Unexpectedly, the compound eyes and ocelli are usually not the receptor sites for photoperiod: the epidermis, the brain or even the cerebral neurosecretory cells themselves may react to the stimuli.

Photoperiod is the major factor inducing diapause in many insects;

temperature is of the greatest importance in its termination. The effect of low temperature in diapause development in *Hyalophora cecropia* has already been described (p. 143). In *Bombyx*, development recommences at 22·5°C if the diapausing eggs experience 80 days at a temperature of 2·5°C, about 65 days at 12·5°C or 160 days at 15°C. Diapause is not broken even after long periods at 0°C or 17·5°C. The temperature of 5°C is therefore optimal for diapause development.

The optimum temperature for diapause development is closely related to the environment in which the insect lives.[67] Species in cold climates usually require low temperatures: in the emperor moth, *Saturnia pavonina*, it lies between −15 and +7°C. Most insects in temperate regions have optima between 0 and 7°C. But in insects which aestivate to avoid hot, dry summers, the optimum temperatures for diapause development are correspondingly high—as much as 35°C for the diapausing eggs of the South African brown locust. In general, those conditions which the diapause is intended to avoid are largely instrumental in bringing about its eventual termination.

Environmental factors which switch off the endocrine system are often quite different from those which switch it on again. In post-embryonic diapauses, the cerebral neurosecretory cells play a central part in the insect's reaction to these exogenous stimuli. Little is known of the internal pathways which transmit these stimuli to the cells (see Chapter 14). It is quite likely that the neurosecretory cells, or neurones in close proximity, react to photoperiod and temperature,[1, 290, 291] although how they can go through a number of moults producing thoracotrophic hormone normally before they shut down is unknown. Still more mysterious is the way in which environmental factors can terminate egg diapause, which often occurs before the embryonic endocrine centres are established. Possibly the egg tissues have a general sensitivity to such stimuli, a sensitivity which becomes confined and limited to special cells in the post-embryonic stages.

HORMONES AND DIURESIS

Terrestrial insects usually have to conserve water, and it is now axiomatic that the cuticular wax layer, spiracular closing mechanisms and uricotelic nitrogen excretion are all adaptations to this end. But circumstances often arise when the insect is presented with an excessive water load—taken in with the food, for example—which must be quickly disposed of if the blood is not to be so diluted that vital functions are impaired.

In blood-sucking insects such as *Rhodnius prolixus*, large quantities of unnecessary fluid are imbibed with each meal, which with other waste materials are excreted through the malpighian tubules. An *in vitro* preparation of malpighian tubules, together with a piece of attached rectum, will function normally for a long time in a physiological saline covered with liquid paraffin. The rectal cuticle has an affinity for the paraffin, and consequently the rectum opens out onto the liquid paraffin layer[199] (Fig. 8.12). Droplets of fluid from the malpighian tubules pass out onto the

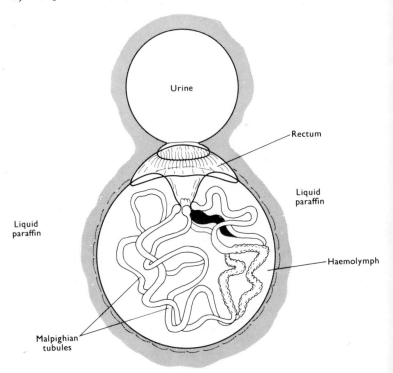

Fig. 8.12 A preparation of malpighian tubules of *Rhodnius prolixus* in liquid paraffin. The tubules, with part of the rectum attached, lie within a bubble of haemolymph in the liquid paraffin. The cuticular lining of the rectum, with an affinity for the paraffin, causes the rectum to spread out along the interface between haemolymph and paraffin. Urine produced by the malpighian tubules forms a droplet where it leaves the rectal lumen. The diameter and hence volume of this droplet can be measured periodically. (After Maddrell[199])

paraffin through the rectum, and therefore the rate of excretion of the malpighian tubules can be measured, together with the total volume of fluid filtered through the tubules (Fig. 8.13).

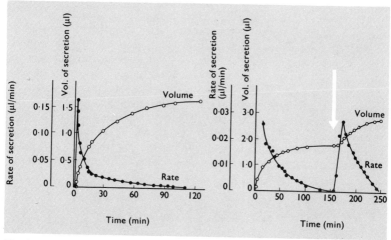

Fig. 8.13 (*left*)   Volume and rate of production of urine produced by a preparation shown in Fig. 8.12.

Fig. 8.14 (*right*)   Volume and rate of production of urine produced when the malpighian tubules are bathed in haemolymph of unfed and fed *Rhodnius*. An arrow marks the time when 'unfed' haemolymph was replaced by 'fed' haemolymph.

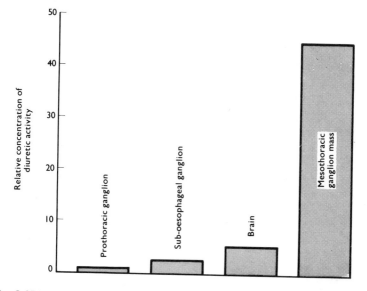

Fig. 8.15   Relative diuretic activity of different parts of the central nervous system of *Rhodnius*.

Malpighian tubules bathed in haemolymph from a fed *Rhodnius* produce a much more copious urine in a shorter time than when they are exposed to haemolymph from an unfed bug (Fig. 8.14). This immediately suggests that diuresis is accelerated by a hormone present in the fed insects, which because it has a positive effect upon urine production, is a diuretic hormone. But extracts of the known hormone-producing tissues in *Rhodnius* have little effect upon malpighian tubule activity (Fig. 8.15); in fact only extracts of the fused thoracic and 1st abdominal ganglia have diuretic activity[199] (Fig. 8.15). All the activity is contained in the posterior part of the ganglionic mass, where large neurosecretory cells are located (Fig. 8.16). In *Rhodnius*, therefore, the diuretic

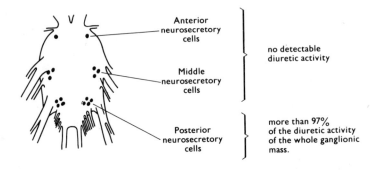

Fig. 8.16 Relative diuretic activity of different parts of the mesothoracic ganglionic mass (containing groups of neurosecretory cells) of *Rhodnius*. (Figs. 8.14–8.16 after Maddrell[199])

hormone is neurosecretory in origin, and has no relationship with the major cerebral endocrine system. Diuresis begins about 30 seconds after *Rhodnius* has fed, so the hormone must be released very rapidly after feeding. Hormone release is inhibited when the ventral nerve cord is cut: it is likely, therefore, that the abdominal expansion caused by feeding, which stimulates stretch receptors whose signals eventually lead to activation of the cerebral endocrine system, also releases diuretic hormone from the ventral ganglionic neurosecretory cells.[199]

Rhodnius, and perhaps all blood-sucking insects, are unusual in that there are periodic crises in water-loading: the diuretic mechanism may be a special adaptation to this manner of feeding. In another bug, the cotton stainer *Dysdercus fasciatus*, a diuretic hormone is produced by the cerebral neurosecretory cells[17] (Fig. 8.17). In the locusts, *Schistocerca gregaria* and *Locusta migratoria*, the cerebral neurosecretory cells also produce a diuretic hormone. Its existence in these locusts was first

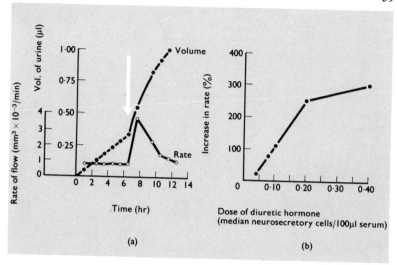

Fig. 8.17 Diuresis in *Dysdercus fasciatus*. (a) The volume and rate of flow of urine from an isolated malpighian tubule preparation (similar to that in Fig. 8.12). The arrow marks the addition of 'diuretic hormone' (0·1 of a group of median neurosecretory cells/100 μl serum). (b) A dose-response curve for 'diuretic hormone' activity. (After Berridge[17])

suspected when the endocrine control of reproduction was being investigated. Destruction of the cerebral neurosecretory cells resulted in a large increase in the haemolymph volume—sometimes to as much as three times its normal value. The abdomens of the operated individuals were so swollen that at first they appeared to be carrying fully developed eggs; only dissection revealed the true reason for the enlargement. Clearly, destruction of the cerebral neurosecretory cells prevented the normal removal of water taken in with the food.[139]

In *Schistocerca*, movement of fluid through the malpighian tubules can be estimated by injecting a neutral dye such as amaranth into the haemolymph. Small samples of blood are taken at regular intervals after injections, and the concentration of amaranth determined by comparison with a known concentration of the dye, for example by means of an absorptiometer. The rate of removal of amaranth from the haemolymph gives a measure of malpighian tubule activity[209] (Fig. 8.18). Extrapolating the graph back to the time of injection of the known quantity of amaranth gives the concentration of dye in the blood at this time—from which the blood volume of the insect can be calculated (Fig. 8.18).

After cautery of the cerebral neurosecretory cells, the activity of the

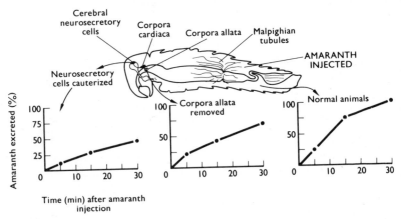

Fig. 8.18 Amaranth clearance from the haemolymph of neurosecretory cell cauterized, allatectomized and normal desert locusts (*Schistocerca gregaria*).

malpighian tubules is low and the blood volume increased (Fig. 8.18). After allatectomy, malpighian tubule activity is reduced, though not so much as in individuals without cerebral neurosecretory cells, and the blood volume is slightly greater than normal (Fig. 8.18). This again suggests an effect of the corpus allatum hormone upon neurosecretory activity (see p. 122): a direct effect of the hormone upon diuresis is unlikely, since extracts of the corpora allata do not affect the rate of removal of amaranth. But extracts of corpora cardiaca injected into locusts with their cerebral neurosecretory cells destroyed have a marked stimulatory effect upon malpighian tubule activity (Fig. 8.20). Increasing concentrations of corpus cardiacum extracts stimulate malpighian tubule activity in a linear manner (Fig. 8.19). Brain extracts also stimulate the malpighian tubules, suggesting that it is the contained neurosecretion of the corpora cardiaca which is the diuretic hormone. In the locust, the intrinsic glandular cells of the corpora cardiaca can be relatively easily separated from the 'neurosecretory storage' parts of the glands (Fig. 2.14). Extracts of the glandular regions have little effect upon diuresis when injected into locusts; extracts of the neurosecretory storage regions have the same effect as extracts of whole glands (Fig. 8.20), confirming that the diuretic hormone originates in the cerebral neurosecretory cells.

When amaranth is injected into the locust some time after the injection of corpus cardiacum extract, it is removed more slowly than when it is injected simultaneously with the extract (Fig. 8.21). This is because some of the hormone has been excreted or destroyed or even used up in

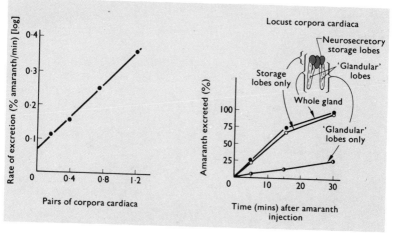

Fig. 8.19 (*left*)   Dose response curve for desert locust corpora cardiaca/amaranth clearance rate in the desert locust.

Fig. 8.20 (*right*)   The effects of storage and glandular regions, and whole corpora cardiaca upon amaranth clearance in the desert locust. The storage lobes possess almost all the diuretic activity of the whole glands.

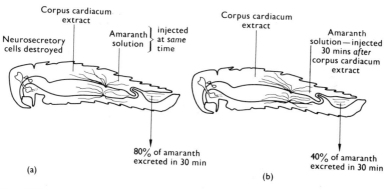

Fig. 8.21   An experiment to demonstrate the biological life of diuretic hormone in the desert locust. The extract loses about half its activity in 30 minutes: its effect upon amaranth clearance when the dye is injected simultaneously with the extract (a) is twice that when the dye is injected 30 minutes after the extract (b).

the period between the two injections. This technique in fact measures the biological life of the hormone in the blood. The diuretic hormone loses about 50% of its activity every hour.[209] Measuring malpighian tubule activity in a previously starved locust at different times after the recommencement of feeding indicates an increasing concentration of

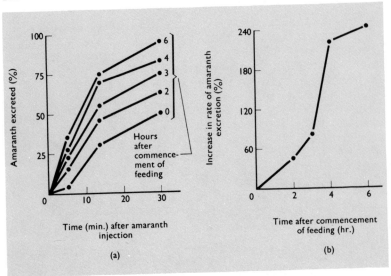

Fig. 8.22 (a) Amaranth clearance rates at different times after feeding previously starved locusts. (b) The percentage increase in the rate of amaranth clearance at these times compared with that in a starved locust. The increasing rates indicate increasing concentrations of diuretic hormone in the blood.

diuretic hormone in the blood (Fig. 8.22). Knowing its biological life, the release of diuretic hormone from the cerebral neurosecretory system can be calculated. This proves to be quite close to the estimates made from histological changes in the neurosecretory system and illustrates again the remarkable effects of feeding upon neurosecretory activity.

The desert locust, like most terrestrial insects, has a very efficient water reabsorbing mechanism in the hind gut, localized particularly in the rectal glands.[223] The diuretic hormone increases the movement of fluid through the malpighian tubules, but this cannot be reabsorbed by the rectum, otherwise the total water content of the insect would remain unchanged. Mordue has developed a technique to measure the effect of the locust diuretic hormone upon rectal reabsorption.[209] The rectum is removed, turned inside out and both its ends ligatured, so that it forms a sac with its cuticle outside. When suspended in saline, water moves into the sac (i.e. from rectal lumen side to haemolymph side of the rectal wall, as it would when reabsorbed from a rectum *in situ*) and the rate of movement can easily be measured by weighing the sac at intervals, the cuticle on the surface being easily dried. When corpus cardiacum extracts are included in the sac, i.e. are placed on the haemolymph side of the rectal wall, water movement into the sac is *inhibited* (Fig. 8.23). The

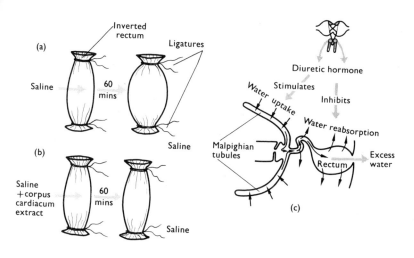

Fig. 8.23 Diuretic hormone and rectal reabsorption in the desert locust. The rectum is turned inside out and its absorption of water measured periodically by the increase in weight of the preparation. Weight increases rapidly when the rectum is filled with saline (a), but is prevented when 'hormone' extract is included (b). (c) shows diagrammatically the dual effects of the locust diuretic hormone, stimulating malpighian tubule activity and inhibiting rectal reabsorption. (Figs. 8.18–8.23 after Mordue[209])

lowest concentrations of corpus cardiacum extracts which stimulate the malpighian tubules have no effect upon rectal reabsorption: the threshold of the response of the rectum is higher than that of the malpighian tubules. So in the desert locust, the diuretic hormone has a double effect: it *stimulates* fluid movement through the malpighian tubules, and inhibits its reabsorption by the rectum. Its net result, therefore, is the loss of water by the insect.

In the cockroach, *Blatta germanica*, and some other insects, destruction of the cerebral neurosecretory cells also causes increases in the haemolymph volume. It is likely that a cerebral diuretic hormone is thus of quite general occurrence.

HORMONES AND PERISTALSIS

In the cockroach *Periplaneta americana*, extracts of the corpora cardiaca will accelerate the rate and increase the amplitude of the heart beat, and increase both gut peristalsis and the twisting movements of the malpighian tubules. The hormone(s) responsible do not act directly, but stimulate pericardial cells in the vicinity of the heart, and analogous cells in the gut, to secrete a tryptamine compound (related to a compound found in the vertebrates: serotonin or 5-hydroxytryptamine) which is the *immediate* cause of the effects[70, 71] (Fig. 8.24). The corpus cardiacum

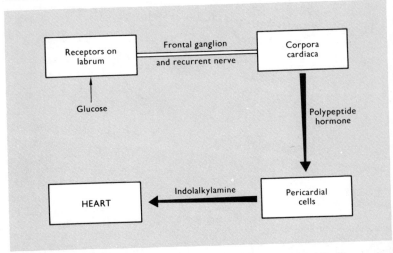

Fig. 8.24 The heart accelerating effect of feeding in the cockroach. The corpora cardiaca release a factor on the receipt of nervous stimuli generated by labral receptors reacting to the presence of glucose. The cardiacal factor indirectly affects heart beat by stimulating the pericardial cells to produce an indolalkylamine. (From data in Davey[70, 71])

hormone is a peptide, and is released when the cockroach feeds. Because of the close juxtaposition of glandular and neurosecretory storage regions in the cockroach, it is impossible to decide in which part the factor originates. In the desert locust, glandular extracts are more potent than storage extracts of the corpora cardiaca in their effects upon the heart[136]: it is possible that two hormones are involved, one from each region. In the locust, there also seems to be some interaction between the storage and glandular regions in the production of the factor: the effects of feeding upon its release tend to support such a suggestion. In any event, the release by feeding of a factor which increases gut movement, the circulation of blood, and malpighian tubule movement is of obvious advantage to the insect in its digestion of food, transport of metabolites, and excretion of waste products.[71, 72]

It is possible that a similar factor is involved in the complex movements of muscles involved in oviposition. In *Rhodnius*, the presence of the cerebral neurosecretory cells are certainly necessary at this time,[75] and in the locust changes in the histology of the cerebral neurosecretory system during egg-laying suggest that neurosecretion is involved[133] (p. 140). But an interaction between neurosecretion and the production of an intrinsic hormone by the cells of the corpora cardiaca cannot be ruled out.

HORMONES AND HYPERGLYCAEMIA

In *Periplaneta*, an extract of the corpora cardiaca elevates the blood sugar concentration of the haemolymph. Two hormones are involved, one very much more potent than the other, and both have a similar effect: to convert the enzyme phosphorylase in the fatbody from its inactive to its active form[251] (Fig. 8.25). The active phosphorylase

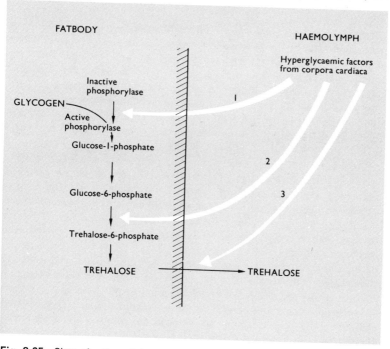

Fig. 8.25 Sites of action of the cockroach hyperglycaemic factors. Two enzyme reactions are affected (1,2) as well as the transfer of trehalose from fat body cells to haemolymph (3). (After Steele[251])

catalyses an important step in the breakdown of glycogen and the net result is an increased production of the disaccharide trehalose (Fig. 8.25).

In the locust, similarly, two hyperglycaemic factors are released from the corpora cardiaca, with the same effect upon the phosphorylase enzyme. In the locust, the ease of separation of neurosecretory storage from glandular regions of the corpora cardiaca allows the particular origins of the two hormones to be determined: the less potent of the two comes from the neurosecretory region, and the more potent from the gland cells. It can be assumed that their origins are similar in cockroaches. But even so, the effects of starvation, feeding, hyperactivity and stress upon the release of *both* hormones in the locust again suggests some interaction between the two parts of the corpora cardiaca. A trophic effect of neurosecretion upon the intrinsic glandular cells of the corpora cardiaca is not at all unlikely.

HORMONES AND TANNING OF THE ADULT CUTICLE

In the blowflies, *Sarcophaga bullata*, *Phormia regina* and *Calliphora erythrocephala*, the median cerebral neurosecretory cells produce a hormone, now called **bursicon**, which hardens and darkens the adult cuticle.[95] Bursicon is a peptide hormone, with a molecular weight of about 40 000. Although produced by the cerebral neurosecretory cells, it is apparently released from the fused thoraco-abdominal ganglia. This alone suggests it is distinct from the thoracotrophic hormone also produced by the cerebral neurosecretory cells; experiment confirms this distinction.

Blood from the newly emerged adults of the cockroach (*Periplaneta americana*), the meal worm (*Tenebrio molitor*), the waxmoth (*Galleria melonella*), the milkweed bug (*Oncopeltus fasciatus*), and the locusts (*Schistocerca gregaria* and *Locusta migratoria*) contains bursicon as shown by injection into newly emerged adult blowflies ligatured tightly behind the head. In the cockroach and the locusts, bursicon is released from the last abdominal ganglion.

In *Calliphora*, cuticular hardening and darkening is delayed until the adult reaches the surface of the substrate in which pupation occurred. The puparium and the pupal cuticle inside are split by a special device on the head of the emerging adult—the ptilinum—which expands and brings local pressure to bear on the external coverings. Thus unlike most other insects, in which the surrounding cuticle of the previous instar is split by expansion of the emerging instar, air swallowing in the blowfly and expansion of the adult is separated by a period of time from moulting proper. A close control over cuticular hardening is clearly necessary.

An endocrine control of the process by something other than that which occurs during juvenile moults (see p. 90*ff*) is necessitated by the degeneration of the equivalent of the thoracic gland during adult emergence.

HORMONES AND BEHAVIOUR

Many behavioural patterns associated with particular developmental events in the lives of insects are influenced by the developmental hormones. Courtship behaviour in female grasshoppers, involving response to the male, sometimes including stridulation, and terminating in copulation, is mediated by the corpus allatum hormone—which is also involved in the control of oocyte development. Cocoon spinning behaviour before pupation in Lepidoptera is eliminated by removal of the cerebral neurosecretory cells. Copulatory behaviour in male mantids is stimulated by a hormone from the corpora cardiaca. But diurnal locomotory behaviour in the cockroach, *Periplaneta americana*, is controlled by neurosecretory cells in the sub-oesophageal ganglion, although cerebral neurosecretion or ecdysone probably overrides this control during moulting, so that the insect remains quiescent.

It is unlikely that hormones affect behaviour in insects through influencing their sensory receptors or effector organs. The available evidence favours the view that integrative centres in the central nervous system are sensitive to hormone concentrations. It is probable that the neural mechanisms for different, or alternative, behaviour patterns already exist in the central nervous system; the hormone would then act to switch one or other of the neural patterns into activity. The mechanisms of endocrine-controlled behaviour in insects are similar to those described from vertebrates. For a fuller treatment of the processes, reference should be made to the reviews of Manning,[201] Highnam[135] and Barth.[8]

NEUROSECRETORY HORMONES IN INSECTS

A multiplicity of effects have been described for hormones from the corpora cardiaca and the cerebral neurosecretory cells. Are these varied effects due to the same hormones, or to a number of different hormones? In some instances, differentiation between hormones can be made. Thus bursicon injected into *Rhodnius* will not activate the thoracic glands and is consequently different from thoracotrophic hormone, although both are produced from the cerebral neurosecretory cells. These cells are composed of a number of different histological types (p. 23) and it can be assumed that they produce different hormones. The diuretic hormone from the neurosecretory storage regions of the locust corpora cardiaca has the same chemical properties as the less potent hyperglycaemic hormone, and when assayed has both diuretic and hyperglycaemic effects. But even so, the two hormones are not necessarily identical: they could be either polypeptides with similar electrophoretic properties, but different constitution, or even different compounds attached to similar polypeptides. An extract from the corpora cardiaca controls lipid mobilization from the fatbody: what relationship this bears to other corpus cardiacum hormones is unknown. The hormone controlling protein synthesis in some insects may be the same as the thoracotrophic hormone, but the evidence is only circumstantial. In the cockroach, one of the hormones controlling heart beat is said to be identical with the thoracotrophic hormone, but this is not necessarily so in the locust.[105, 106] Until purification and identification of insect neurosecretory hormones becomes possible, many of these problems will remain unsolved. The difficulties exposed by attempts to characterize the neurosecretory hormones from the median eminence of vertebrates are likely to be exacerbated by the relatively small size of insects and the consequent minute amounts of the hormones able to be extracted. The identification of the thoracotrophic hormone in *Bombyx* (p. 214*ff*) has shown already what problems there are.

9

Endocrine Mechanisms in Crustacea—I

Many developmental and physiological processes in Crustacea are controlled by hormones, but often the exact mechanisms are confused. This is due to a combination of factors, such as the great difficulty in maintaining large numbers of animals in controlled conditions in the laboratory, the long life cycles of some of the large species which are popular for experimentation, and a frequent lack of detailed information about the normal habitat and behaviour of the animals. Furthermore, almost all the observations on crustacean endocrine mechanisms refer to the Malacostraca, and mainly to the Decapoda at that. All this must be borne in mind when general conclusions are drawn about the endocrine control of moulting, sexual differentiation and reproduction, colour change, migration of retinal pigments and heart beat in Crustacea.[41]

HORMONES AND MOULTING

The nature of the moult cycle

Information about the nature of the moult cycle and its control is limited to the Decapoda, and in fact often refers specifically only to the crabs (Brachyura). Knowledge of the process in other groups of Crustacea is fragmentary.

In the Crustacea, as in insects, moulting is a part of the mechanism of growth. Change in form and increase in size can only occur when the hard calcareous exoskeleton is shed and before the new cuticle is hardened. Periodic ecdysis, by splitting the old cuticle through expansion of the new instar, brought about by taking up water through the digestive tract, is as characteristic of the crustacean as it is of the aquatic insect. The increase in size and weight of the crustacean during ecdysis does not constitute growth. This must be defined as the increase in dry weight of the body which occurs in the periods between moults, when the absorbed water is gradually replaced by protein. Thus although ecdysis, increase in size and increase in total weight are all markedly discontinuous in Crustacea, growth itself is a continuous process (Fig. 9.1), as it is in insects (Fig. 6.19) and almost all other animals.

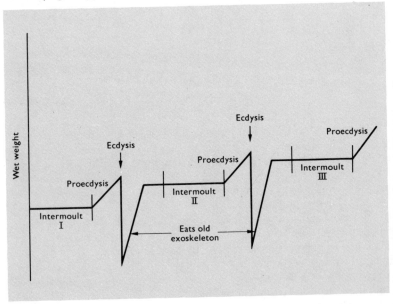

Fig. 9.1   Changes in wet weight during the growth of a crab. Growth is apparently discontinuous: water is taken up during proecdysis resulting in a weight increase. The shedding of the old cuticle causes a sudden decrease in weight, but this is regained in many crabs which eat their cast skins. Although the wet weight remains constant after the moult, growth actually occurs during this period, the water taken up at proecdysis being replaced by protein.

Moulting dominates the life of the crustacean. Ecdysis cannot be considered as a brief interruption of the normal life of the crustacean, but rather as a process which has far reaching effects upon the whole physiology. In this respect, many decapods differ profoundly from the pterygote insects in that moulting proceeds throughout adult life. In fact, very little is known about the control of larval growth and moulting in the decapods: when the endocrine control of growth and moulting in Crustacea and insects is compared, it must always be remembered that in the former the process refers almost exclusively to postmetamorphic stages, whereas in the latter metamorphosis is usually an intrinsic part of the mechanism.

Post-metamorphic development in Crustacea follows one of several patterns. The shore crab, *Carcinus maenas*, moults several times after metamorphosis before the puberty moult, at which the sexual appendages appear and the animal becomes sexually mature and is able to reproduce; moulting then continues for up to 3 years after the puberty moult. In the spider crab, *Maia squinado*, on the contrary, the puberty moult is the final one, and the animal therefore never moults after becoming

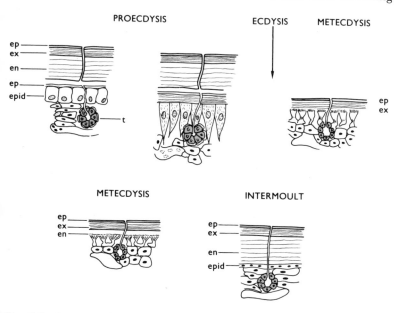

Fig. 9.2 Structure of the integument of the land crab, *Gecarcinus lateralis*, at each stage of the moult cycle. During early proecdysis the epidermal cells enlarge, separate from the old cuticle (apolysis), and secrete a new epicuticle. By late proecdysis the epidermal cells have enlarged still further and secretion of the new exocuticle has begun. After ecdysis, exocuticle secretion is complete and endocuticle production begins, to be completed during metecdysis. The epidermal cells then decrease in size and remain small during the intermoult period. *Abbreviations:* en, endocuticle; *ep*, epicuticle; *ex*, exocuticle; *epid*, epidermal cells; *t*, tegumental gland. (After Carlisle and Knowles[41])

sexually mature. In other decapods, such as the edible crab, *Cancer pagurus*, and the lobster, *Homarus vulgaris*, growth and moulting appear to continue until death.[41]

The moulting cycle can be divided into four stages (Fig. 9.2) which facilitates analysis of its control.[40] No sharp divisions exist between these stages.

Stage 1. Proecdysis

This is the preparation for moulting. The first signs of proecdysis are activation of the epidermal cells and hepatopancreas. The epidermal cells separate from the cuticle, a process known as **apolysis**, and then divide. Almost immediately the epidermal cells begin to secrete the new exoskeleton. At the same time calcium is removed from the old cuticle, resulting in an increased blood calcium concentration. As these processes continue, the animal stops feeding and becomes inactive: during this time the reserves of the hepatopancreas are utilized. Splitting of the old cuticle marks the end of proecdysis.

Stage 2. Ecdysis

This is the short period during which the animal sheds the remains of the old cuticle. There is a rapid uptake of water. The animal does not feed.

Stage 3. Metecdysis

This begins with the newly moulted animal, its exoskeleton still soft and extensible as water uptake continues. Mineral deposition begins in the exocuticle, and later endocuticle secretion starts. Initially, the animal still does not feed, continuing to utilize reserves in the hepatopancreas. But during the latter half of metecdysis feeding recommences, the production of the exoskeleton is completed, and tissue growth occurs, replacing the absorbed water. Both protein and DNA have high turnover rates during this time, and the tissues double their dry mass, losing water in direct proportion.

Stage 4. Intermoult

Both skeletal formation and tissue growth have been completed, but feeding continues and metabolites in excess of current requirement are stored in the hepatopancreas. Lipid is the major reserve, but some glycogen and protein are also stored. This intermoult period is often referred to as the period of normality, but the specific accumulation of reserves in preparation for the next moult is no more normal than any other part of the moult cycle.

If the animal moults seasonally, the intermoult period is long, and is known as **anecdysis**; if moults are much more frequent, occurring throughout the year, the period is called **diecdysis**, hardly distinguishable as metecdysis merges gradually into proecdysis.

Endocrine control of moulting[218]

The initiation of proecdysis

Over 60 years ago, it was discovered that removal of the eyestalks during the intermoult period accelerated ecdysis and initiated precocious growth. But the significance of these effects was not realized until later work on colour change in Crustacea (p. 197ff) revealed the presence of the eyestalk neurosecretory system (Fig. 2.19).

Subsequent experiments showed that removal of the eyestalks during the intermoult period results in the premature onset of proecdysis, but their removal during proecdysis has no effect. These results suggested that the eyestalks are the source of a hormone which normally *inhibits* the onset of proecdysis. But the converse of this experiment, the implantation of the source of the hormone into eyestalkless animals, initially yielded variable results: sometimes proecdysis was delayed, at other times there was no effect. The criteria for demonstrating a hormonal effect (p. 17) were not always fulfilled.[18] In these experiments, sinus glands were implanted as the probable source of the hormone; it was not realized at this time that the glands are neurohaemal organs, storing perhaps only small quantities of hormone produced elsewhere. But when the true nature of the sinus glands was recognized, whole eyestalk neurosecretory systems were implanted into eyestalkless animals, and the accelerated onset of proecdysis was delayed.[217] Later, the selective destruction of either the sinus gland, or the X-organ, confirmed that the eyestalk neurosecretory system produces a moult-inhibiting hormone.

But it was soon discovered that the neurosecretory moult-inhibiting hormone did not act directly upon the tissues to delay proecdysis. When the glandular paired Y-organs (p. 24) were removed from *Carcinus* during the intermoult period, or very early proecdysis, ecdysis was prevented. Removal of the Y-organs later in proecdysis had no effect upon the subsequent moult, but the *next* ecdysis was blocked. Reimplantation of Y-organs into such a blocked *Carcinus* initiated the moult cycle again. And when supernumerary Y-organs were implanted into normal animals, the intermoult period was shortened. These results all point to the production by the Y-organs of a moulting hormone, with a positive effect upon the tissues.[99, 82, 83] The crustacean Y-organs are thus analogous to the insect thoracic glands: both produce a moulting hormone. But whereas the neurosecretory thoracotrophic hormone of insects *stimulates* the thoracic glands into activity in preparation for the moult, the neurosecretory moult-inhibiting hormone of Crustacea *inhibits* the activity of the Y-organs during the intermoult period, its absence before proecdysis allowing the Y-organs to secrete (Fig. 9.3). But in spite of this funda-

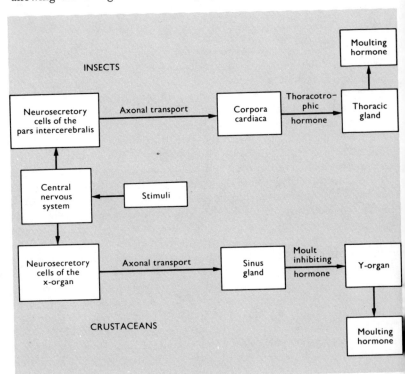

Fig. 9.3 Control of moulting in crustaceans and insects. In both groups moulting is controlled by a two-step hormonal sequence: the moulting hormone is secreted by an epithelial endocrine gland whose activity is controlled by a neurosecretory hormone. In crustaceans, the neurosecretory hormone is produced in the X-organs and released from the sinus glands and *inhibits* the activity of the Y-organs: moulting hormone is produced when secretion of the moult inhibiting hormone ceases. In insects, the neurosecretory hormone from the pars intercerebralis of the brain is released from the corpora cardiaca and *stimulates* the thoracic glands to produce the moulting hormone.

mental difference, in both Crustacea and insects moulting is controlled by a two-step endocrine sequence: control over an epithelial endocrine gland exercised by neurosecretory hormones, and control over the body tissues exercised by hormones from the epithelial glands.

There is little information available on the fate of the Y-organs in those crustaceans which cease to moult, but in *Maia* the Y-organs atrophy when the animal reaches maturity. *Maia* is thus similar to the insects where the thoracic glands degenerate after the final moult.

Other hormones in the moult cycle

The endocrine control of moulting in Crustacea, outlined above, rests upon a firm experimental foundation. But other hormones have been claimed to be involved in the moult cycle. In the prawn, *Leander serratus*, eyestalk removal from individuals in some populations results in the delayed onset of proecdysis; in other populations proecdysis is brought forward, the intermoult period being shortened, when the eyestalks are removed. The latter situation closely parallels that in *Carcinus*, the operation removing the source of the moult inhibitory hormone. But *delayed* proecdysis following eyestalk removal can only mean that the source of a moult *accelerating* hormone has been removed.[36]

The moult-accelerating hormone is said to be produced by neurosecretory cells in the brain and other parts of the central nervous system. In the Brachyura and Astacura, the source of the moult-accelerating hormone is unaffected by eyestalk removal which takes away only the moult-inhibitory hormone: consequently the Y-organs are stimulated and proecdysis is brought forward. But in those species which have the sensory pore X-organ as a neurohaemal organ in addition to the sinus gland, the medulla terminalis ganglionic X-organ (Fig. 2.20) is said to be an *additional* source of the moult-accelerating hormone. Eyestalk removal in these species therefore either accelerates or delays proecdysis according to the prevailing hormone balance at the time of the operation. It must be admitted that although such a complicated control over Y-organ activity is feasible, the evidence for its existence is not good. The insects have certainly developed a simpler, energy conserving system.

There is better evidence for the existence of a hormone, or hormones, regulating the uptake and retention of water at ecdysis. A normal *Carcinus* increases in volume by about 80% at each moult; the increase is nearer 180% in eyestalkless animals, due entirely to a greater water content. Sinus gland extracts injected into eyestalkless animals counteract the abnormal water uptake, and when injected into normal *Carcinus*, less water than usual is taken up at the moult.[38]

The water-balance hormone is distinct from the moult inhibitory hormone. Sinus gland extracts taken from *Carcinus* at any stage of the moult cycle reduce water uptake in both normal and eyestalkless crabs, but the extracts delay proecdysis only when prepared from sinus glands taken during the intermoult period. Clearly, it is possible for the sinus glands to contain water balance hormone without any concomitant moult inhibitory hormone activity. Furthermore, decapods typically shed their exoskeletons at a time and place of their own choosing: ecdysis can be postponed if conditions are unsuitable. The moment of ecdysis is determined by the uptake of water to cause swelling and splitting of the old cuticle. Since this moment can be varied independently of the progress of proecdysis, it supplies further support for the independence of water balance and moult-inhibiting hormones.

The water balance hormone is clearly stored in and released from the sinus glands, but the neurosecretory centre responsible for its manufacture has not yet been determined. The way in which the hormone acts is also unclear: it could act by regulating nervous centres responsible for drinking, or by affecting directly the antennal and maxillary excretory organs.

Eyestalk removal in *Carcinus* thus has a dual effect: the onset of proecdysis is accelerated and water uptake at ecdysis is increased. Several moults may follow the operation, and the considerably shortened intermoult periods do not allow the complete replacement of water by new tissues. Consequently, the eyestalkless animal becomes more water-logged at successive moults with the net result that the animal dies from a dilution of the tissues.

The control of water balance is not the same in all decapods. In the land crab, *Gecarcinus lateralis*, there is no eyestalk hormone which inhibits the proecdysial retention of water—similar amounts of water accumulate in both normal and eyestalkless individuals. It is claimed that the normal proecdysial increase in water content results from the presence of an anti-diuretic hormone. The nature and source of this hormone are unknown, but it *could* be identical with the Y-organ moulting hormone, or at least associated with it in some way.

In *Gecarcinus*, a *diuretic* hormone may also control water balance by getting rid at ecdysis of water retained during proecdysis. Implantation of the fused thoracic ganglia, or of the eyestalk ganglia, into eyestalkless *Gecarcinus* results in water loss.

The evidence in support of the control of water balance by two antagonistic hormones is still very tentative. But such a system would give much more precise control over water content than a single hormone system (Fig. 9.4). *Gecarcinus* is well adapted to terrestrial life on an island, where even dew may contain wind-blown sea salt, and is able to utilize water with salinities ranging from 0% (demineralized distilled

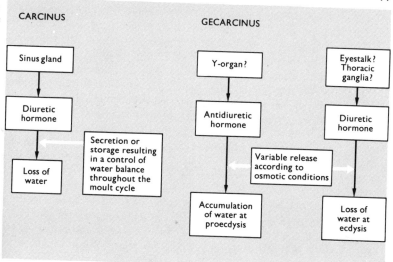

Fig. 9.4 Control of water balance in *Carcinus* and *Gecarcinus*. In *Carcinus*, water balance is controlled by the variable secretion of a single diuretic hormone, but in *Gecarcinus* the control is likely to be exercised by antagonistic diuretic and antidiuretic hormones, although much of the evidence is still rather tentative. The two-hormone system in *Gecarcinus* would result in a more precise control over water balance and could be associated with the island habitat of the crab and the consequent need to utilize water of greatly varying salinities.

water) to 30% (sea water). At intermediate salinities, excessive amounts of water are retained during proecdysis, and a dual control system would be of considerable advantage.[20] Whether such a system is a special adaptation to terrestrial life, or whether decapods other than *Gecarcinus* also possess it is at present entirely speculative.

Mode of action of the moulting hormone

When moulting hormone from the Y-organs is released into the blood to initiate proecdysis, a complicated sequence of biochemical and morphological changes are set in train. It is not yet possible to decide which of these changes are the result of the direct action of the moulting hormone, and which are secondary or tertiary effects, or are even more remotely connected with the hormone's primary action. It is even possible, of course, that the hormone may act on several different processes, or at several points in the same biosynthetic pathway. The Y-organs may even secrete more than one hormone, each having diverse effects.

The Y-organs must be present for the first part of proecdysis, and thereafter moulting can proceed normally in their absence (p. 173). It is sometimes claimed that this indicates a single primary action of the moulting hormone upon its target organs initiating the subsequent changes associated with proecdysis. This effect of the moulting hormone in Crustacea would then be akin to that of ecdysone in the insects (p. 96). But nothing is known of the blood concentrations of moulting hormone during proecdysis: an initially high concentration followed by a steady decrease could act sequentially upon different stages of the moulting cycle.

Uncertainty about the primary action, or actions, of the moulting hormone makes a clear analysis of its control of the moult cycle impossible. But it is still worthwhile to discuss the gross physiological changes which occur during the moult, and to attempt to correlate such changes with moulting hormone activity. The following description refers specifically to *Carcinus maenas*, except where stated otherwise.

An accumulation of glycogen in the epidermal cells and an increased concentration of blood calcium, probably released from the hepatopancreas, and later from the old cuticle, mark the first appearance of moulting hormone in the blood. These early events can be slowed down or even reversed if external conditions become unsuitable, or if the Y-organs are removed. But once the epidermal cells enlarge and separate from the old cuticle, and use their accumulated glycogen to synthesize chitin for the new cuticle, the progress of proecdysis is irreversible.

RESPIRATORY METABOLISM   As the epidermal cells secrete layers of protein and chitin in the new cuticle, their oxygen consumption rises from 0·49 to 0·85 μl O_2/mg dry weight/hour. It is sometimes claimed that the moulting hormone primarily affects respiratory metabolism. After removal of the eyestalks in *Gecarcinus*, the respiratory quotient (R.Q.) falls from 0·77 to 0·69, suggesting a metabolic shift towards lipid utilization. Subsequent implantation of sinus glands restores the R.Q. to its original value.[19] But there is no reason to suppose that these changes in respiratory metabolism are not secondary to some other effect of the moulting hormone; they may reflect an increased synthetic activity associated with the morphological and physiological changes that take place during proecdysis. The low R.Q. for example, could result from the utilization of lipids stored in the hepatopancreas, necessitated by the cessation of feeding at this time.

CALCIUM METABOLISM   Likewise, it is difficult to decide whether calcium metabolism is directly regulated by the moulting hormone. During the latter half of proecdysis, calcium is removed from the old cuticle: it is commonly hoarded in the hepatopancreas, gastroliths (calcium deposits in the stomach lining) or the blood in terrestrial an freshwater Crustacea, and may be either stored or excreted in marine forms. Subsequently, calcium is deposited in the new cuticle, coming initially from the calcium stores, but finally from the external medium since the new cuticle is larger and more massive than the old. Proecdysial calcium resorption and its postecdysial deposition both occur after the Y-organs have ceased to be essential for moulting; thus they cannot be *directly* related to the primary action of the moulting hormone.

Removal of the eyestalks results in the rapid depletion of calcium in the hepatopancreas, and a single injection of Y-organ extract causes a threefold increase in the blood calcium concentration.[39] Both events are characteristic of early proecdysis, but it is not clear what the function of the mobilized calcium is at such an early stage of the moult cycle. Very much more information is required about calcium levels in the blood, hepatopancreas, epidermis and exoskeleton in normal and experimental animals before definite conclusions can be reached.

METABOLISM AND METABOLIC RESERVES The distinctively cyclic nature of the moulting process is reflected by cyclic changes in the reserves of carbohydrates, lipid, proteins and inorganic materials, some of which are not directly concerned with moulting as such. It is possible to relate a few changes to the presence or absence of hormones. In the porcellanid crab, *Petrolisthes cinctipes*, removal of the eyestalk neurosecretory system in the intermoult period accelerates protein mobilization from the hepatopancreas, such as occurs normally in proecdysis. The rates of incorporation of radioactively labelled amino acids into protein suggests that in the hepatopancreas protein synthesis decreases, and in the epidermis and muscles it increases, both in proecdysis and after ablation of the eyestalk neurosecretory system.[179] In the intermoult period, therefore, protein synthesis could be restrained by eyestalk neurosecretory hormones either directly or through their inhibition of the Y-organs. But once more, the results of such experiments do not point exclusively to a direct action of hormones upon protein synthesis.

There are, however, two instances in which there is good experimental evidence for a direct hormonal control of a metabolic process. The first is the regulation of blood sugar concentrations; the second, control over the production of digestive enzymes.

THE HYPERGLYCAEMIC HORMONE Aqueous eyestalk extracts injected into crabs very quickly increase the concentration of reducing sugars in the blood (hyperglycaemia). As little as 0·001 'eyestalk equivalents' doubles the reducing sugar concentration in the blood of *Callinectes sapidus* one hour after injection. Eyestalk removal decreases the reducing sugar level only in some species, but injections of sinus gland extracts induce hyperglycaemia in all species tested.[238]

Stress of any kind also leads to hyperglycaemia, except after eyestalk removal or when the X-organ-sinus gland tracts have been cut. Release of neurosecretory hormone is a characteristic response to stress in many animals. The hyperglycaemic hormone very probably acts upon the hepatopancreas, which is an organ of digestion, storage and intermediary metabolism analogous in many respects to the vertebrate liver and the insect fatbody.

Another aspect of the endocrine control of carbohydrate metabolism has been studied in the freshwater crayfish, *Orconectes virilis*, where carbohydrate utilization in the hepatopancreas varies with the stage of the moulting cycle. During the intermoult period, the hepatopancreas synthesizes pentose sugars, which are vital constituents of nucleotides and nucleic acids. Under normal conditions, this function of the hepatopancreas has little significance in energy production.

Just before proecdysis, carbohydrate metabolism in the hepatopancreas shifts from the pentose cycle to the Embden-Meyerhof glycolytic system. It must be emphasized that *both* pathways always coexist: the difference between the intermoult and proecdysial periods lies in the *relative* balance of importance of the two systems.

When crayfish eyestalk extracts from intermoult animals are injected into premoult individuals, the pentose cycle of carbohydrate metabolism predominates in the hepatopancreas, representing a change back from the normal premoult situation. Conversely, eyestalk extracts from premoult individuals injected into intermoult animals result in a decreased pentose pathway utilization of sugar, so that glycolytic conversion predominates.[204, 205]

Clearly, the eyestalk extracts act by switching carbohydrate metabolism into one of two alternative pathways at the glucose-6-phosphate level (Fig. 9.5). It has been suggested that two eyestalk hormones are involved: the first, predominating in the intermoult period, influences the synthesis or activation of the enzyme glucose-6-phosphate dehydrogenase, thus directing carbohydrate metabolism into the pentose cycle; the other, predominating in the proecdysial period, acts to divert glucose-6-phosphate into the glycolytic direction.

Such a two-hormone system seems clumsy and too complicated for what seems the relatively simple task of controlling specifically the activity of one enzyme, glucose-6-phosphate dehydrogenase. If a high level of activity of the enzyme were maintained during the intermoult period by a single hormone which was absent during proecdysis, the same effect upon carbohydrate metabolism would be achieved. The moult-inhibiting hormone would be a likely candidate. But this system does not explain how proecdysial eyestalk extracts, when injected into intermoult animals, switch carbohydrate metabolism to the glycolytic

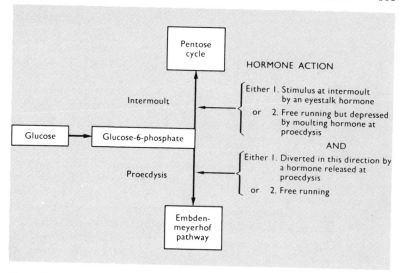

Fig. 9.5 Endocrine control of carbohydrate metabolism in the hepatopancreas of the crayfish, *Orconectes virilis*. Two possible pathways of carbohydrate metabolism exist in the hepatopancreas: *via* the pentose cycle or the Embden-Meyerhof glycolytic system. During the intermoult period, the pentose cycle pathway predominates but just before proecdysis the emphasis changes to glycolysis. There are two ways in which such a change could occur: (1) One eyestalk hormone secreted during the intermoult period might stimulate the synthesis of glucose-6-phosphate dehydrogenase, so directing glucose-6-phosphate into the pentose cycle; another hormone, released at proecdysis, then diverts glucose-6-phosphate into the glycolytic pathway. (2) Both systems could be free-running, with the pentose cycle normally predominating. The moulting hormone, released at proecdysis, might then depress glucose-6-phosphate dehydrogenase activity so allowing glycolysis to predominate.

pathway. Similarly, the pentose cycle could be free running in the intermoult period but the activity of glucose-6-phosphate dehydrogenase depressed by moulting hormone when this appears in proecdysis. If intermoult eyestalks inhibited Y-organ activity, and proecdysial eyestalks stimulated the glands, cross-transplantations of eyestalks between these periods would produce the observed effects, but only because the Y-organs were inhibited or stimulated. When the hepatopancreas is incubated *in vitro* with eyestalk extracts from either intermoult or proecdysial crayfish, no differences in carbohydrate metabolism occur—supporting the suggestion that some other factor is involved. It is not clear what relationship these changes have with the hyperglycaemia produced by the injection of eyestalk extracts, nor is it clear what is the functional significance of these changes.

CONTROL OVER DIGESTIVE ENZYME PRODUCTION   The hepatopancreas produces enzymes to digest fats, proteins and carbohydrates. Eyestalk removal in the crayfish *Procambarus clarki*, results in a decrease in the synthesis of amylase by the hepatopancreas. Like all enzymes, amylase is a protein and consequently its reduced synthesis could reflect an over-all reduction in the basic protein synthetic mechanisms of the cells of the hepatopancreas. In fact, during 7 days from eyestalk removal, the RNA content of the hepatopancreas decreases to zero, to reappear when eye-stalk extracts are subsequently injected.[91] Clearly, a hormone present in the eyestalk must act upon some fundamental metabolic activities of the cells of the hepatopancreas but at what level is as problematical as that of ecdysone upon insect cells (p. 102). Since all RNA disappears after eyestalk removal, presumably none of the enzymes produced by the hepatopancreas are synthesized. Those other than amylase have not yet been investigated, but the cells of the hepatopancreas show degenerative changes in the endoplasmic reticulum after eyestalk removal, which certainly suggests a general effect upon protein synthesis. If removal of the eyestalk does indeed depress hepatopancreas protein synthesis generally, then considerable doubt must be thrown upon the validity of any studies of hepatopancreas metabolism after removal of the eyestalks.

The hormonal control of metabolic processes in Crustacea suffers from insufficient knowledge about the animals' normal metabolism, and also from the lack of pure hormone preparations. The use of whole eyestalk, or even sinus gland, extracts mean that several hormones are injected simultaneously; eyestalk removal similarly removes the source of not one but many hormones. Further, in normal animals, the differential release of separate hormones must occur: injections of whole eyestalks, or sinus glands, may very well provide a mixture of hormones in pro-portions that are never found naturally. Detailed and convincing studies on the control of metabolism in Crustacea must await the availability of pure hormone preparations.

10

Endocrine Mechanisms in Crustacea—II

Endocrine Control of Reproduction

In the insects, sexual differentiation is held to be entirely genetic, both the primary and secondary sex characters developing under the direct influence of the genome without the intervention of circulating hormones. Recent work on the control of sexuality in the glow worm, *Lampyris noctiluca* (p. 103), indicates that such genetic control may not be of general occurrence. But in the Crustacea both sexual differentiation and gonadal activity can be influenced by hormones, to some extent resembling the situation in the vertebrates.

THE CONTROL OF SEXUAL DIFFERENTIATION

Most of the Malacostraca are bisexual, only a few groups showing protandric hermaphroditism. In the bisexual species, sex is determined genetically, but the morphological and functional expression of this genetic sex is largely influenced by hormones. The young malacostracan acquires its specific form either at hatching or at the end of a larval period: in either case, the sexes cannot usually be distinguished and this sexuality may persist for several moults. When sexual differentiation does begin, it progresses at successive moults, and in many species the full development of secondary sexual characters is not complete until well after gonadal maturity. External differences between the sexes may be due not only to the presence of special sex organs, such as the male's penis or the female's oostegites, but also to variations in the morphology of structures common to both (Fig. 10.1).

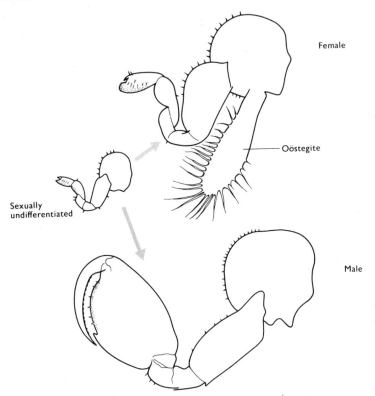

Fig. 10.1 Sexual dimorphism of the second gnathopod of *Orchestia gamma-rellus*. The undifferentiated gnathopod develops into a powerful clawed structure in the male and a slender non-clawed limb in the female. The female gnathopods also develop oostegites—plates which form a brood pouch for the young. (After Charniaux-Cotton[48])

In undifferentiated young Crustacea, the primordia of the vasa deferentia are present in both sexes. Attached near the end of each vas deferens is a gland called, because of its function, the androgenic gland (see below), lying between the muscles of the coxopodites of the last pair of walking legs (Figs. 2.19, 10.3). In a genetic female, the androgenic glands fail to develop, but in the male, the glands enlarge to form solid strands of cells, folded several times (Figs. 10.2, 10.3). The cells of the androgenic gland secrete in a holocrine manner, emptying their total contents into the blood. The androgenic glands were first discovered in the amphipod, *Orchestia gammarellus*, in 1954, and have since been described in representatives of most of the malacostracan orders.[46, 47, 49] How are the androgenic glands involved in the control of sexual differentiation?

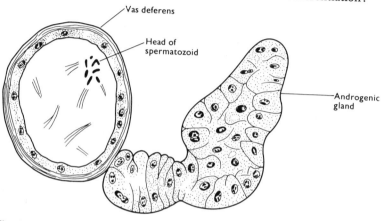

Fig. 10.2 Section through the vas deferens and androgenic gland of *Orchestia gammarellus*. (After Charniaux-Cotton[48])

In *Orchestia*, when the androgenic glands are removed from a young male, the primary germ cells in the testis produce not spermatocytes but oocytes. Usually, this oogenesis is subsequently reversed because it is difficult to remove the glands completely from young animals and they regenerate. But it seems that a hormone from the androgenic glands is necessary for the normal development of the testes; in its absence, the primary germ cells develop spontaneously into oocytes.

In genetic females, the androgenic glands do not develop. But when androgenic glands from a young male are transplanted into a female, the ovaries rapidly transform into testes (Fig. 10.4), the primary germ cells giving rise to spermatocytes, which develop fully through spermatids to spermatozoa. The follicle cells turn into cells similar to those found in the normal testis, which secrete mucus around the spermatozoa. The rudimentary vasa deferentia in the female also grow and develop. [50] The implantation into females of testes, or of vasa deferentia without androgenic glands, has no effect upon ovarian development, but masculinization of the ovaries can be achieved by extracts of androgenic glands injected into females and by blood transfused from a male *Carcinus* into a

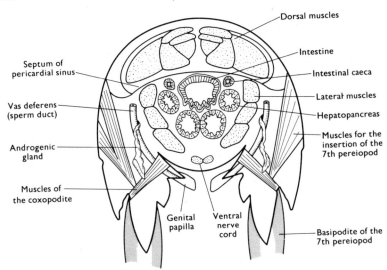

Fig. 10.3 Transverse section through the last thoracic segment of *Orchestia gammarellus*, showing the position of the androgenic glands. (After Charniaux-Cotton[50])

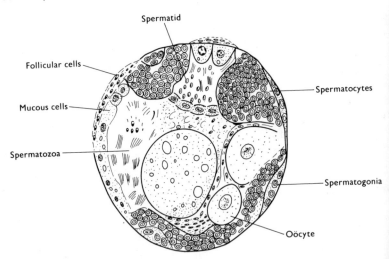

Fig. 10.4 Ovary of a mature female *Orchestia gammarellus* which has been masculinized by the implantation of an androgenic gland. The primary germ cells have given rise to spermatocytes; spermatids and spermatozoa are also present. (After Charniaux-Cotton[50])

female *Orchestia*. The androgenic hormone can act in the female as it does in the male. Moreover, the hormone is not species-specific.

Thus the genetic determination of the primary sex characters—the testis and ovary—is expressed through the androgenic gland. In a genetic male, the glands develop and their androgenic hormone induces the gonads to become normal testes. In a genetic female, the androgenic

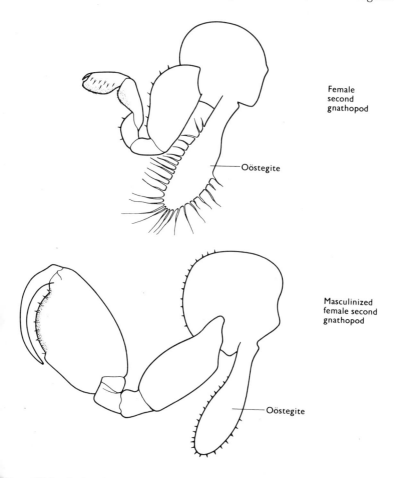

Female
second
gnathopod

——Oöstegite

Masculinized
female second
gnathopod

——Oöstegite

Fig. 10.5 *Orchestia gammarellus :* transformation of the female second gnathopod into the male form after implantation of an androgenic gland. The gnathopod progressively acquires the male form during the period following implantation of the gland. The only evidence of its original female structure is the presence of the oostegite. (After Charniaux-Cotton[48])

glands do not develop, and in the consequent absence of the androgenic hormone, the gonads develop into normal ovaries.

But how do the secondary sexual characteristics develop? Are they determined genetically, or are they also hormonally controlled? In *Orchestia*, the claw on the gnathopod is small and slender in the female, well developed and powerful in the male. In a female whose ovaries have been masculinized by the implantation of androgenic glands, the feminine gnathopod claw is transformed into the masculine form[48] (Fig. 10.5). Even in a female ovariectomized before the androgenic glands are implanted, the female claw transforms to the male type, showing that the effect is due directly to the action of androgenic hormone, and is not mediated through the masculinized gonad. Moreover, females masculinized by the implantation of androgenic gland *behave* exactly like males, and will even mate with normal females. Fertilization is impossible, however, because the newly induced sperm ducts do not have a lumen; but freshly laid eggs can be fertilized when sperm taken from masculinized ovaries is deposited upon them.

The male secondary sex characters are induced by the direct action of androgenic hormone upon the tissues. When the androgenic glands are removed from a male *Orchestia*, and at the same time the appendages are amputated, new appendages regenerate without the typical male form, as would be expected. But the new appendages do not have the female form either—they are undifferentiated rather than being male or female (Fig. 10.6). This can only mean that in the female, where the androgenic

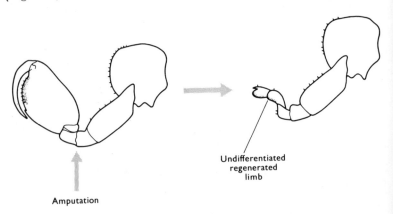

Undifferentiated
regenerated
limb

Amputation

Fig. 10.6 Regeneration of the second gnathopod of a male *Orchestia gammarellus* when the androgenic gland is removed at the same time that the limb is amputated. The gnathopod regenerates as a sexually undifferentiated limb. (After Charniaux-Cotton[48])

glands are undeveloped, the secondary sex characters must be determined by something other than the mere absence of androgenic hormone.

The female *Orchestia* possesses oostegites—plates which form a brood pouch for the young—which is a permanent secondary sex character. But in addition, long ovigerous hairs appear on the oostegites during a moult preceding the laying of eggs, replacing the short juvenile hairs of immature forms, or of females in sexual repose [48] (Fig. 10.7).

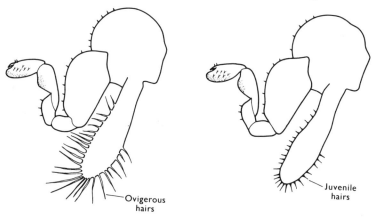

Fig. 10.7 Second gnathopods of female *Orchestia gammarellus* showing the *juvenile* hairs on the oostegites of immature animals or those in sexual repose; and the *ovigerous* hairs which appear on the oostegites during the moult preceding egg laying.

When an ovary from a young or a maturing female is implanted into a male whose androgenic glands have been removed, oostegites appear at the first or second postoperative moult. This permanent female secondary sex character is thus induced by an ovarian hormone. Similarly, total ovariectomy of a reproductive female results in the replacement of ovigerous by juvenile hairs at the first or second postoperative moult. When an ovary, or any portion of one, is implanted into an ovariectomized female, yolk deposition in the oocytes begins after the first postoperative moult, and ovigerous hairs are developed after the second postoperative moult. Thus the temporary female sex characters are also controlled by an ovarian hormone. [45]

It is likely that in *Orchestia*, the ovarian hormones controlling oostegite development and the formation of ovigerous hairs are distinct from each other (Fig. 10.8). The latter is secreted only when vitellogenesis proceeds, which accounts for the intermittency of ovigerous hair development, and since all parts of the ovary produce the hormone the follicle cells are its

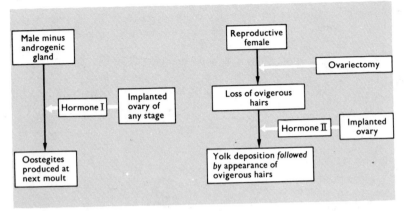

Fig. 10.8 Experiments which suggest that two hormones are produced by the ovary, controlling oostegite development and ovigerous hair formation respectively. An ovary implanted into a male whose androgenic glands have been removed will always induce oostegite formation at the next moult, implying a constant secretion of this hormone. The ovarian hormone inducing ovigerous hair formation is secreted *only* when vitellogenesis proceeds.

probable source. But the oostegite-controlling hormone must be secreted continuously, to explain the invariable effect of ovary transplantation into males without androgenic glands. The source of this hormone is probably the germinative zone or young follicles.

PARASITIC CASTRATION[51]

In the Crustacea, protandric hermaphroditism is readily explained by a hormonal mechanism of sexual differentiation. The regression of the androgenic glands terminates the male phase and the testes are transformed to ovaries in the absence of androgenic hormone. In the same way, the frequent occurrence of intersexes in the Malacostraca is understandable: a partial failure in the hormonal mechanism of control can lead to gonads which contain spermatozoa and oocytes, and to mixtures of male and female secondary characters. Testes containing some oocytes are common, due to failure of some germinative cells to get sufficient or any androgenic hormone.

The phenomenon of parasitic castration is also more readily explained now that it is known that sexual differentiation is hormonally controlled. Male crustacea parasitized by *Sacculina carcini* or other Rhizocephala, show changes in their secondary sex characters, which often resemble the female form. It was previously thought that destruction of the testes brought about these changes, or that the parasite imposed metabolic demands upon the male which were similar to those caused by developing eggs in the female, and the male was feminized as a result.

In fact, there is no castration of the male by the parasite, although infrequently there may be slight atrophy of the testes. Moreover, the secondary sex characters of the female crustacean appear before vitellogenesis (see above) and femaleness therefore cannot be due to excessive metabolic demands. But very significantly, the testes of parasitized males frequently contain oocytes, and rarely the testes are transformed completely into ovaries. Consequently, parasitic castration must now be explained in terms of a reduction in the amount of androgenic hormone in the blood. It is likely that the parasite removes the hormone from the blood. This is followed by transient hypertrophy of the androgenic glands, in an attempt to produce the high concentration of hormone necessary to maintain the male characters. Later the glands atrophy, probably as the result of their previous unnatural hyperactivity together with the influence of the parasite. The male sex characters regress more or less completely at each moult as the androgenic hormone disappears; the appearance and development of female characters at subsequent moults would then depend upon the degree of metamorphosis of testes into ovaries.

THE CONTROL OF GONADAL ACTIVITY

Reproductive activity and growth in decapod Crustacea are antagonistic, as they are in the polychaete worms (p. 52). Ovigerous shrimps and crabs will not moult until after the eggs have hatched. In the shrimp *Palaemon serratus*, the ovaries are quiescent during the summer, but vitellogenesis proceeds rapidly in October, and the first spawning occurs in November. Every moult is then followed by egg laying until the following May or June. But when the eyestalks are removed from *Palaemon* during the quiescent summer period, vitellogenesis begins at once. After the subsequent moult, the brooding characteristics appear and the eggs are laid. The implantation of eyestalk extracts into eyestalk-less females prevents the precocious vitellogenesis.[215] Similar effects follow these operations in many other shrimps and crabs.[37, 116] They suggest that a hormone inhibiting vitellogenesis is produced by the X-organ-sinus gland system in decapod Crustacea (Fig. 10.9). The X-organs show histological cycles associated with oocyte development,[4] and in some isopods ovariectomy causes hypertrophy of the sinus glands; this provides supporting evidence for a relationship between the ovaries and the eyestalk neurosecretory system.

Accelerated vitellogenesis in maturing eyestalkless *Palaemon* is not merely a consequence of a reduced intermoult period and precocious proecdysis which would follow removal of the moult-inhibiting hormone (p. 173). In fact, premature vitellogenesis *prolongs* the intermoult period in the same manner as in a normal female. If the eyestalks are removed from an immature female *Palaemon*, the oocytes develop no more than they do in control animals, but somatic growth and ecdysis are accelerated. The ovarian inhibiting hormone thus appears not to be identical with the moult-inhibiting hormone. In juvenile forms, somatic growth and ecdysis take precedence over ovarian development; in mature individuals reproductive growth is of greater importance. Thus according to

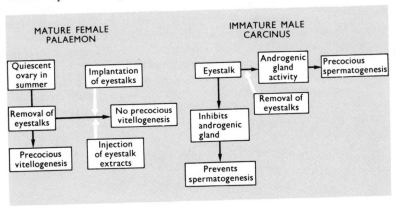

Fig. 10.9 Control of gonadal activity in decapod crustaceans. In the mature female, the eyestalk produces a hormone inhibiting vitellogenesis. In the immature male, spermatogenesis is similarly hormonally inhibited, but the eyestalk hormone acts by inhibiting the activity of the androgenic gland.

the developmental stage of an individual, eyestalk removal will accelerate whichever process is currently dominant.

The control of ovarian activity is further complicated by recent reports that the implantation of a brain, or the fused thoracic ganglia, into the crab *Parathelphusa guerini*, results in ovarian enlargement. This raises the possibility of a bi-hormonal control of ovarian development, as for somatic growth (p. 175). The details of such a control are equally obscure.

In young male *Carcinus*, eyestalk removal causes precocious spermatogenesis, and growth of the vasa deferentia.[77] That these effects are produced by removing inhibition from the androgenic glands is very likely, since the glands hypertrophy within a month after the operation (Fig. 10.9). Eyestalk removal in mature crabs has no effect upon the activity of the testes or the androgenic glands. Inhibitory hormones thus control spermatogenesis in the male and vitellogenesis in the female, but the male hormone differs profoundly from that of the female in that it acts through epithelial endocrine organs, the androgenic glands, and at a quite different stage in the life-history.

In protandrous hermaphrodites, the change from the male to the female phase is accompanied by the degeneration of the androgenic glands (p. 190). In the prawn *Lysmata seticaudata*, eyestalk extracts injected into males of the size at which sex reversal normally takes place considerably reduce the degree of change, whereas eyestalk removal in such animals enhances the degree of change.[37] Thus the eyestalk neurosecretory system may maintain the male phase in some animals but only at the time during which sex normally changes. The exact timing of sex reversal could be correlated with external conditions through such a mechanism.

II

Endocrine Mechanisms in
Crustacea—III

In the Crustacea, neurosecretory hormones are of the utmost importance in controlling growth and development, both through their trophic effects upon epithelial endocrine glands (Y-organs, androgenic glands) and also by a direct intervention in metabolic processes. But neurosecretory hormones are also concerned in the regulation of much more temporary physiological events, such as heart beat, colour change and the movement of retinal pigments, which are related to short-term environmental fluctuations (Chapter 14).

HORMONAL CONTROL OF HEART BEAT

The crustacean heart is essentially a single-chambered sac of striated muscle, pierced by afferent and efferent openings. In most Crustacea, the heart is neurogenic, with the elongated dorsal cardiac ganglion as pacemaker. Inhibitory and acceleratory nerve fibres can affect the beating of the heart, acting directly upon the neurones of the cardiac ganglion. In the decapods, one inhibitory and two acceleratory fibres enter the nerve plexus, which forms part of the pericardial organ (Fig. 11.2) on each side of the heart, before passing to the cardiac ganglion as a single dorsal nerve[203] (Fig. 11.1).

In decapod Crustacea, major portions of the pericardial organs lie across the openings of the branchio-cardiac veins (Fig. 11.2), and

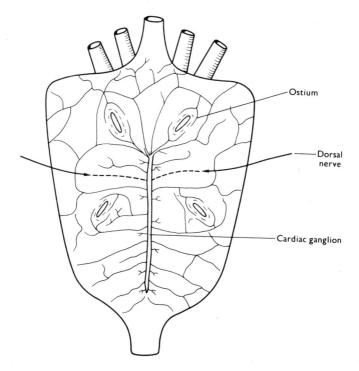

Fig. 11.1 Dorsal view of the heart of the lobster showing its innervation. The single dorsal nerve which pierces the wall of the heart on each side and enters the cardiac ganglion consists of both inhibitory and acceleratory nerve fibres. (After Maynard[203])

elementary neurosecretory granules, about 1200 Å in diameter, are present within the nerves making up the organs. The position of the pericardial organs across the flow of venous blood from the gills, together with their content of neurosecretion, suggests a role in regulating the heart beat.[12]

When isolated hearts have aqueous extracts of pericardial organs added to the perfusion medium, the amplitude of beat is always increased, and often the frequency of beating increases also. Since the heart is isolated from its nervous connections, the increased amplitude and frequency of beating cannot be due to any effect upon the inhibitory or excitatory fibres which run to the cardiac ganglion. But when nerve fibres from the cardiac ganglion are monitored to record their nervous impulses, both the frequency and duration of bursts of impulses increase when the ganglion is treated with aqueous extracts of pericardial organs[64] (Fig. 11.3).

Electrophoresis and chromatography of extracts of pericardial organs of *Cancer* separate two peptides with cardio-excitor activity. Both peptides have a similar amino acid composition. Whether these represent two distinct hormones, or are merely the degradation products of a single hormone, is conjectural.[12] It had previously been thought that the decapod cardio-excitor was 5-hydroxytryptamine or 5:6-dihydroxytryptamine. But responses of the heart to 5-HT can be blocked without affecting its reaction to aqueous extracts of pericardial organs or to the separated peptides[64] (Fig. 11.4). This suggests that the pericardial organs

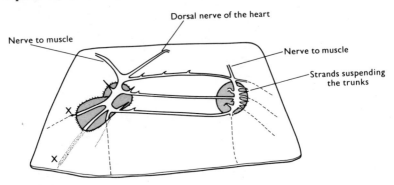

Fig. 11.2 Pericardial organs of the right side of the spider crab, *Maia squinado*. These are neurohaemal organs lying within the pericardial cavity and bathed in blood moving towards the heart. The organs are shown in place on the inside of the lateral pericardial wall. The nerves from the central nervous system which enter the pericardial organs are shown as dotted lines, and the points at which the two anterior nerves enter the lumen of the veins are shown as crosses. (After Carlisle and Knowles[41])

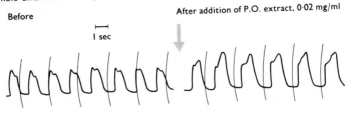

Fig. 11.3 A record of the electrical activity from an isolated cardiac ganglion of the lobster. The ganglion is held in the perfusion fluid by forceps, anteriorly and posteriorly, which also serve as recording electrodes. When extract of pericardial organs is added to the perfusion fluid, the burst *duration* increases. The burst *rate* may increase or decrease when extract is added. (After Cooke[64])

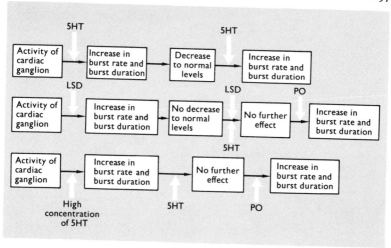

Fig. 11.4 The action of pericardial organ extract (PO), 5-hydroxytryptamine (5HT) and D-lysergic acid diethylamide (LSD) upon the electrical activity of the isolated, perfused cardiac ganglion of the lobster. LSD blocks the responsiveness of the ganglion to 5HT but not to PO. This suggests that the cardio-excitor substance from the pericardial organs is not 5-hydroxytryptamine.

release a peptide neurosecretory product, designed to initiate the action of a neurotransmitter acting upon the neurones of the cardiac ganglion.

It might be wondered why hormonal cardio-excitors are necessary when the crustacean heart can be influenced by *nervous* inhibitors and accelerators. The answer probably lies in the different natures of the two control mechanisms: the effects of the nervous regulators are essentially transitory, while that of the hormone can be very much more prolonged.

HORMONAL CONTROL OF COLOUR CHANGE[5, 88]

For more than a hundred years biologists have attempted to explain the spectacular changes in colour shown by crustaceans. Colour change may be slow and predictable, synchronized with cyclic environmental events such as season, the ebb and flow of tides, or the daily progression of night and day; or it may be quickly and reversibly associated with local and transient fluctuations in illumination or background colour.

The effectors which bring about colour changes in an individual are the chromatophores, situated directly beneath the epidermis or scattered in the deeper tissues of the body. The crustacean chromatophore is a much branched cell, whose overall shape remains unchanged during colour changes. Instead, pigment contained within each chromatophore disperses or concentrates to alter its shade or tint (Fig. 11.5). When dark pigment is dispersed through each of many chromatophores, an overall dark colouration of the animal results; when the pigment is concentrated, the animal appears blanched.

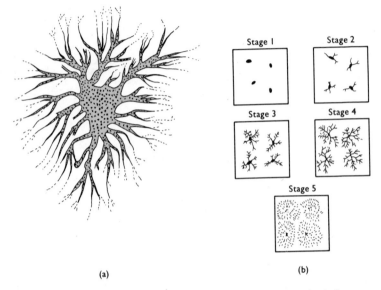

(a)   (b)

Fig. 11.5 (a) Diagram of a chromatophore from the uropod of the prawn, *Leander serratus*; the pigment is shown in its expanded state. (b) The arbitrary scale by which measurements of pigment expansion in chromatophores are made. In Stage 1 the pigment is fully contracted; Stages 2, 3 and 4 indicate progressive pigment expansion; Stage 5 is a chromatophore with pigment fully expanded.

Crustacean chromatophores may contain one (monochromatic), two (dichromatic) or several (polychromatic) pigments. Some chromatophores are multinucleate, and seem to be groups of anastomosing cells. A polychromatic chromatophore might be a syncytium of physiologically distinct cells, each containing a single pigment. Red, blue, yellow, white and black pigments are commonly found in crustacean chromatophores. Polychromatic chromatophores are characteristic of the prawns and shrimps (Decapoda Natantia); mono- and dichromatic chromatophores are found in the decapod Astacura, Anomura and Brachyura, and also in the Isopoda and Stomatopoda. Most experimental work on colour change has been limited to the Decapoda.

The shrimp, *Crangon crangon*, has a typical polychromatic chromatophore pattern with four types of pigment cell: monochromatic black, dichromatic black-red, trichromatic brown-yellow-red, and tetrachromatic black-white-yellow-red. The fiddler crab, *Uca pugilator*, a brachyuran, has three kinds of monochromatic chromatophores, containing red, white or black pigment.

The control of polychromatic chromatophore systems is likely to be complex, for potentially the *independent* movement of several pigments is involved. At one time, pigment movements were thought to be under nervous control in spite of the failure of fifty years of experimentation to provide evidence for the hypothesis. But in 1928, it was found that the blood of the shrimp *Crangon crangon*, and the prawn *Palaemonetes vulgaris*, contained substances which affected the movements of chromatophores.[222, 176] This provided the first indisputable evidence for the existence of a hormone in any invertebrate, and the endocrine control of colour changes in the Crustacea has been intensively studied ever since.

The source of chromactivating hormones

After the discovery of chromactivating hormones in the blood, a systematic search suggested that tissues near the basement membrane of the eyes were a potent source of the hormones. Histologically, two possible secretory organs, the sinus gland and the X-organs (p. 25) were found in this region. Soon after, chromactivating substances were also found in the central nervous system outside the eyestalks, the circumoesophageal connectives and the post-oesophageal commissures being particularly potent sources for the materials.

It was only in 1951, when the concept of neurosecretion had become firmly established, that the full significance of chromactivating hormones originating in the central nervous system was realized. It is now known that the crustacean eyestalk X-organ-sinus gland system is the source of many neurosecretory hormones, some of which control development and reproduction (Chapters 9, 10). But the chromactivating hormones in the circum-oesophageal and post-oesophageal commissures are also neurosecretory hormones in transit from other neurosecretory centres to neurohaemal organs in the epineurium of the posterior commissure nerves—the post-commissural organs (Fig. 2.19, and p. 25).

The control of pigment movement

Once eyestalks and other parts of the central nervous system had been identified as sources of chromactivating hormones—now called ***chromatophorotrophins***—it was quickly shown that individual chromatophores would react to more than one hormone. In *Crangon*, one hormone, soluble in alcohol, from the post-commissural organs concentrates the pigment in black chromatophores on the body, causing an overall blanching; another, insoluble in alcohol, also from the post-commissural organs, disperses the pigment in the black chromatophores on the body *and on the tail*, making the individual an overall dark shade. But in addition, the eyestalks of *Crangon* also contain two hormones, one of which blanches the body only, the other blanching the body and the tail, by concentrating the pigment in the black chromatophores. In *Crangon*, therefore, as many as four hormones are involved in controlling the movement of pigment in one chromatophore type[28] (Fig. 11.6). But

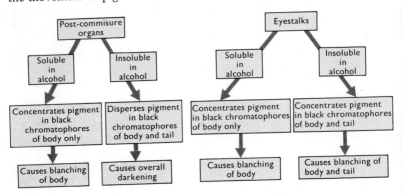

Fig. 11.6 Hormones controlling the movement of the black pigment in the chromatophores of *Crangon crangon*. Alcoholic and aqueous extracts of the post-commissure organs and the eyestalks suggest the presence of four different chromatophorotrophins controlling pigment movement in this one chromatophore type.

one very important principle emerges: when a chromatophorotrophin induces either dispersion or concentration of chromatophore pigment, its *absence* alone does not bring about the opposite movement. This is caused by a second hormone which acts antagonistically to the first. Such antagonistic dispersing and concentrating chromatophorotrophins have now been identified in at least eight crustacean genera, and the 'multiple hormone hypothesis' for chromatophore control is well established, although the *complete* control system for all chromatophores in any one species has not yet been fully worked out.

In the prawn, *Palaemonetes*, the sinus glands contain a hormone which concentrates pigment in the red chromatophores, and another, probably identical, in the tritocerebral commissures, and also in the circum-oeso-phageal commissures. An antagonistic red pigment dispersing chromatophorotrophin is present also in the circum-oesophageal commissures and in the abdominal nerve cord.[26] The presence of the dispersing chromatophorotrophin is difficult to detect. If normal prawns are kept on a white background, the red pigment is fully contracted, but injections of nerve cord extracts containing dispersing hormone have only a slight and transient effect. But when eyestalkless animals (thus with a major source of concentrating hormone removed) are kept on a dark background, the red chromatophores are fully expanded and the injection of sinus gland extracts (containing concentrating hormone) into these animals causes rapid and complete concentration. After a little time, the effect of the

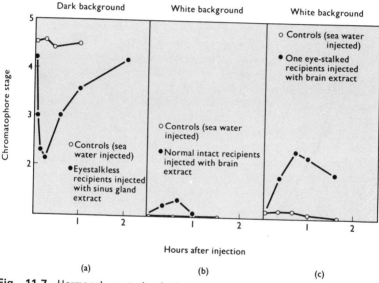

Fig. 11.7 Hormonal control of the red chromatophores of the prawn, *Palaemonetes vulgaris.* (a) Animals kept on a dark background have maximally dispersed red chromatophores. The injection of extracts of the sinus glands causes an immediate and almost complete concentration of the red pigment. (b) Animals kept on a white background have maximally concentrated red chromatophores. The injection of an extract of the brain causes only a slight and temporary dispersion of the red pigment. These experiments show that two antagonistic hormones control the movement of red pigment, and that the concentrating hormone is more potent than the dispersing one.

(c) But if one eyestalk is removed, the effectiveness of the dispersing hormone is greatly increased, presumably because the amount of concentrating hormone in the blood has been reduced.

The antagonism between the pigment dispersing and concentrating hormones is facilitated when a large amount of one hormone acts during the time that the concentration of the second hormone is diminishing. (After Brown, Webb and Sandeen[26])

hormone wears off, and the pigment starts to expand: injection of extracts of dispersing hormone at this time markedly accelerates the rate of pigment dispersion (Fig. 11.7).

This experiment shows that the red pigment dispersing hormone of *Palaemonetes* is less potent than the concentrating hormone. But it also illustrates an important facet of the ***two hormone hypothesis***, namely, that antagonism between the hormones is achieved normally by a high concentration of one hormone acting when the concentration of the second is falling. Removing only one eyestalk from *Palaemonetes* makes more effective the action of a quantity of dispersing hormone which is, ineffective when injected into intact animals, again because the titre of concentrating hormone has been reduced, at least for a time.

The presence of chromatophorotrophins throughout the central nervous system of Crustacea poses the problem of whether they can properly be called hormones, since one half of the essential endocrinological experiment—removal of the source of the hormone—is impossible, for the animals die as soon as large parts of the nervous system are removed. Only chromatophorotrophins from the eyestalk neurosecretory system would be true hormones if the strict letter of the definition were followed. But the injection of extracts of the nervous system has a hormonal effect, and it is possible to show, even in normal animals, that their chromatophorotrophins are released into the blood. This has been done in *Palaemonetes* by keeping the animals for long periods on either a white or a black background. When transferred to the opposite background, the chromatophores take longer to change to the opposite condition the longer they have been maintained under the previous conditions (Fig. 11.8). This result is interpreted to mean that a greater concentration of dispersing or concentrating hormone is maintained *in the blood* when particular conditions are prolonged. But even more significantly, after 14 days on a white background, the central nervous system contains very much more dispersing hormone when assayed on eyestalkless animals than it does after only 2 hours. Similarly, the sinus glands, circumoesophageal and tritocerebral commissures contain *less* concentrating hormone after 14 days on a white background than after 2 hours, when similarly assayed. Exactly opposite effects are found when the animals are kept on a black background for 14 days.[92] These results can only mean that the hormone which is not in use during these prolonged periods accumulates within the nervous system. Together with the evidence for high blood levels of the antagonistic hormone, this provides extremely good evidence for the origin of chromatophorotrophins within the nervous system, *and* their release into the blood. Although circumstantially, all the chromatophorotrophins are therefore hormones.

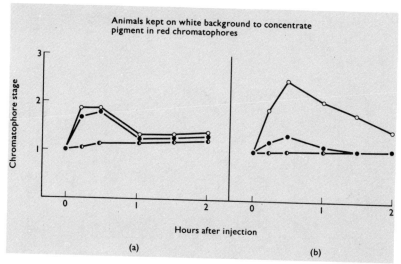

Animals kept on white background to concentrate pigment in red chromatophores

Hours after injection

(a)　　　　　　　(b)

Fig. 11.8 Response of the red chromatophores of the prawn, *Palaemonetes vulgaris*, to extracts of circum-oesophageal connectives. (a) Animals kept on a white background received injections of extract of circum-oesophageal connectives from animals that had been kept on a white background (○) or a black background (●) for two hours. Control animals (◑) received injections of sea water. (b) Animals again on a white background received injections of extract of circum-oesophageal connectives from animals kept on a white background (○) or a black background (●) for two weeks. Control animals (◑) received injections of sea water.

The results show that the amount of red pigment dispersing hormone in the circum-oesophageal connectives increases significantly when animals are kept for long periods on a white background. So when the red pigment in the chromatophores is maintained in a concentrated state, dispersing hormone is not released and accumulates in the circum-oesophageal connectives. (After Fingerman, Sandeen and Lowe[92])

Diurnal colour change

Uca pugilator, the fiddler crab, changes colour in a daily rhythm, becoming dark by day and pale by night. The rhythm persists for a long time when the crab is kept in constant darkness; and when the crabs are kept at 0–3°C for several hours, the rhythm is delayed by an interval closely approximating the period of chilling (Fig. 11.9). The rhythm must therefore be controlled internally, and does not merely reflect changes in environmental illumination. The rhythm is expressed mainly through pigment movements in the monochromatic black chromatophores which are predominant in *Uca*, although red, white and yellow chromatophores are also present.[23]

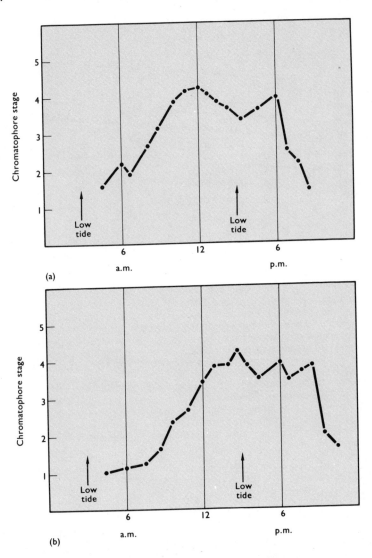

Fig. 11.9 The daily rhythm of melanin dispersion in the fiddler crab, *Uca pugnax*. (a) the normal rhythm which is maintained in the laboratory after the crabs are collected. (b) The rhythm in crabs collected at the same time as those in (a), but placed at 0–3°C for three hours. The rhythm is interrupted by the low temperature period, and the maximum dispersion within the chromatophores is delayed by three hours. (After Brown, Fingerman, Sandeen and Webb[23])

When the eyestalks of *Uca* are removed, the animals become permanently pale due to concentration of pigment in the black chromatophores. The eyestalks must be the source of a black pigment dispersing hormone. When sinus glands are implanted into eyestalkless *Uca*, the rhythm is restored. It must be supposed that the pigment dispersing hormone from an isolated implanted sinus gland diffuses constantly out of the gland into the blood. The restoration of the rhythm must therefore result from the nightly release of an antagonistic pigment concentrating hormone from a source other than the sinus gland (Fig. 11.10). This is supported by the

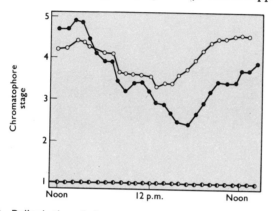

Fig. 11.10 Daily rhythm of pigment movement in the black chromatophores of normal and operated *Uca pugnax*. ○—normal crabs. ●—eyestalkless crabs that have received an implant of three sinus glands. ◑—eyestalkless crabs. For further details see text. (After Fingerman [89]).

fact that extracts of sinus glands are less effective in dispersing the black pigment in eyestalkless *Uca* when injected at night than when injected during the day.[89]

Uca shows a tidal rhythm of pigment movement superimposed upon the diurnal rhythm. Supplementary dispersion of melanin in the chromatophore occurs between 1–3 hours prior to the time of low tide. Thus when low tide occurs near midday, the peak of diurnal chromatophore pigment movement is reinforced, whereas low tides in the early morning and evening cause a double peak of pigment dispersion (Fig. 11.11). The two rhythms are functionally related: when the diurnal rhythm is experimentally shifted by exposing *Uca* to light during the normal dark period, the tidal rhythm of pigment dispersion is also shifted. Rather similar rhythms of colour change occur also in other crabs.

When extracts of *Uca* sinus glands are separated electrophoretically and the fractions tested for pigment dispersing activity on isolated legs

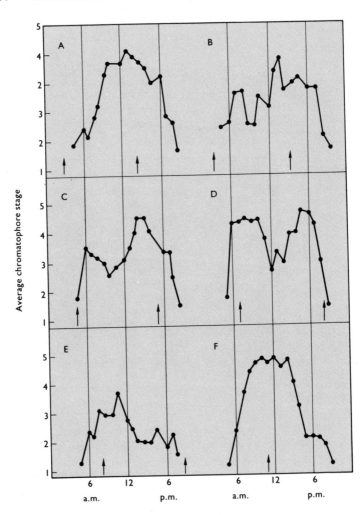

Fig. 11.11 Changes in melanin dispersion on the day following collection in *Uca* kept in darkness. The arrows indicate the times of low tides on the days the observations were made. There are clearly two cycles of melanin dispersion: one is a diurnal rhythm, and the other a tidal rhythm with greatest dispersion of melanin occurring 1–3 hours prior to the time of low tide. Thus when low tide occurs near midday, the peak of melanin dispersion is reinforced, whereas low tide in the early morning or evening results in a double peak of melanin dispersion because the two rhythms are out of phase. (After Brown, Fingerman, Sandeen and Webb[23])

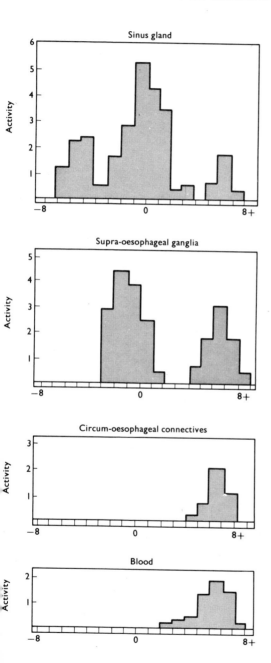

Fig. 11.12 Electrophoretic separation of melanin dispersing substances from the sinus glands, the supra-oesophageal ganglion, the circum-oesophageal connectives and the blood of *Uca*. Aqueous extracts of the tissues are made and separated by paper electrophoresis. After electrophoresis, the papers are cut into strips, the material on the strips eluted, and the eluates tested on an isolated leg of *Uca* with maximally concentrated chromatophores. The sinus glands produce three pigment dispersing substances, the supra-oesophageal ganglion two, and the circum-oesophageal connectives and the blood one. That found in the supra-oesophageal ganglion and the sinus glands, but not in the blood, may be a precursor of the active hormone released into the blood. (After Fingerman [89])

with maximally concentrated chromatophores, three substances are found. Brain extracts similarly are found to contain two pigment dispersing substances, with the same electrophoretic mobility as two of the sinus gland substances. The circum-oesophageal commissures and the blood contain only one pigment dispersing substance, with an electrophoretic mobility similar to the third sinus gland substance and unlike the two brain substances[89] (Fig. 11.12). The simplest explanation of these results is that the two brain substances and the corresponding fractions in the sinus gland are precursors of the one hormone found in the blood, some of which, naturally, is also found in the sinus gland.

When the same separations are carried out, but the fractions assayed on an isolated leg with maximally *expanded* chromatophores, only the circum-oesophageal connectives are found to contain a black pigment concentrating substance. This has the same electrophoretic mobility as the dispersing substance in the sinus gland, so if it were present in the sinus gland, it would be masked by the dispersing hormone.[89] But the experiments described above suggest that it is released, in quantity at least, at a site other than the sinus gland.

Significance of colour change

In general, chromatophores are of considerable importance for protective colouration, thermoregulation and displays associated with mating and parental behaviour. Because its colour merges with the sea bottom, an individual is protected against predation from above; blanching or transparency camouflages with the sea's surface and protects against predation from below. It is not surprising, therefore, that direct light and reflected light together control the particular shade of the individual. It is likely that the ommatidia in particular parts of the compound eyes, acting through the central nervous system, control the release of dispersing and concentrating hormones (Fig. 11.13). The presence of antagonistic pairs of chromatophorotrophins in Crustacea may thus be related ultimately to integration centres in the central nervous system which effect the release of the appropriate hormones.

In addition, pigmentation of the body can be used for thermoregulation. In *Uca*, the black pigment in the chromatophores tends to concentrate when the temperature rises above 15°C. The white pigment tends to concentrate at a temperature above 20°C.[25] The concentration of the black pigment and expansion of the white with increasing temperature has the effect of reducing the dark areas which absorb light and heat, and increasing the reflective white areas. If the intensity of illumination increases, the white chromatophores of three species of *Palaemonetes* expand even when the animals are on a black background. But in

Palaemonetes, unlike *Uca*, the white pigment concentrates with increasing temperature. Heat and intense light usually occur together in nature, as in direct sunlight, and the antagonistic responses of the white

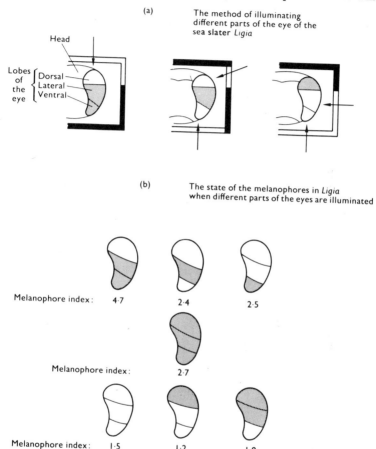

Fig. 11.13 Effects of illuminating different parts of the compound eyes upon the state of the chromatophores in *Ligia*. (a) Method used to illuminate different parts of the eye. (b) Relationship between illumination and the chromatophore index. When the dorsal part of the eye is illuminated, as would happen naturally with direct light upon a dark background, the chromatophore pigment disperses fully, darkening the animal to tone with its surroundings. When the eye is illuminated from the ventral surface, or from several directions, as would happen with direct light together with light reflected from the background, the chromatophore pigment concentrates, toning the animal with a light background.

8

pigment in *Palaemonetes* to the two factors may be a device to keep the chromatophores in a steady state.[93] But only the combination of detailed ecological and physiological studies is likely to reveal the manifold value of chromatophores to the animal in its natural habitat.

Control of retinal pigment movement

The crustacean compound eye, like that of the insect, consists of a number of units or ommatidia. Each ommatidium is surrounded by two sets of pigment cells—distal and proximal—and in the vicinity of their

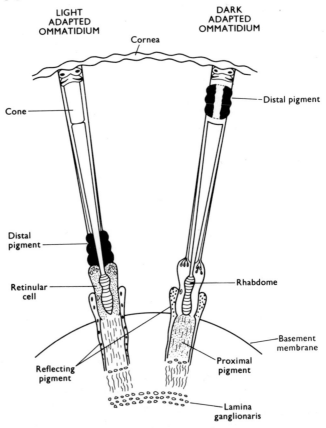

Fig. 11.14 Position of pigments around an ommatidium of a crustacean eye in the light adapted (left) and the dark adapted (right) condition. The compound eyes, are composed of large numbers of such ommatidia. (After Carlisle and Knowles[41])

basement membranes, cells containing 'reflecting' pigment are present.

In the light adapted eye, the distal and proximal pigments prevent the sensitive rhabdome from receiving too much light, and the reflecting pigment cells move beneath the basement membrane (Fig. 11.14). In the dark adapted eye, the distal and proximal segments uncover the rhabdome, and the reflecting pigment moves externally to the basement membrane (Fig. 11.14). The rhabdomes thus receive light from adjacent as well as their own corneas.[170]

When extracts of eyestalk are injected into *Palaemonetes*, the distal pigment migrates towards the light-adapted position; sinus gland and brain extracts have the same effect.[171]

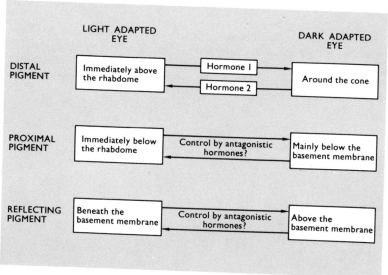

Fig. 11.15 Hormonal control of pigment movement in the crustacean eye. Movement of the distal pigment is controlled by two antagonistic hormones. It is not yet certain that movement of the proximal and reflecting pigments is hormonally controlled. For further details see text.

Sinus gland extracts injected into an animal maintained in a light intensity such that the distal pigment is midway between the fully dark adapted and light adapted positions, causes a light adapting response which lasts for 2–3½ hours, but this is followed by a dark adapting response which lasts for 3½–6 hours. Does the sinus gland therefore contain two antagonistic hormones or is the dark adapting response merely over-compensation following the disappearance of the light adapting hormone?

The injection of brain extracts produces a light adapting response equal to that produced by the sinus glands, but there is no subsequent dark adapting response. Electrophoretic separation of sinus gland extracts produces a fraction which induces a light adapting response after injection, and another with a dark adapting effect. The movement of the distal retinal pigment is thus controlled by two antagonistic hormones (Fig. 11.15). Moreover, when *Palaemonetes* is kept in the dark for long periods of time, extracts of brain can have up to four times the normal concentration of light adapting hormone, suggesting its accumulation when conditions do not call for its release.[24, 22, 27]

The proximal and reflecting pigments also appear to be hormonally controlled, but the evidence is meagre and confused, and will not be discussed further here.

Electrophoretic separation of eyestalk extracts of *Cambarellus shufeldtii*, produces distinct fractions with light adapting, red pigment dispersing and red pigment concentrating effects.[87] The hormones controlling retinal pigment migration are therefore likely to be quite separate from the chromatophorotrophins (p. 218).

12

The Chemistry of Invertebrate Hormones

Only in the insects and crustaceans is the chemical nature of a few hormones known with any degree of certainty, so this chapter must necessarily be limited to these two groups. In the invertebrates generally, hormones which are clearly the products of neurosecretory cells are difficult to isolate and purify, and little detailed work on their chemical characterization has yet been attempted.

Before a hormone can be analysed chemically it must be extracted from the animal in pure form and in sufficient quantity. This entails its separation from other materials by partition through a variety of solvents and its eventual crystallization. At every stage of the separation and purification procedure, the biological activity of the extracts has to be tested to ensure that the hormone has passed into one solvent and has been removed from another. The extraction and purification of a hormone therefore involves the development of a suitable bio-assay, which for preference should be simple, quick and easily repeatable so that accurate statistical estimates of activity can be made. The assay used for the eventual identification of ecdysone and juvenile hormone in insects both come into this category; the crustacean chromatophorotrophins, which can often be assayed on isolated pieces of cuticle, are another good example.

NEUROSECRETORY HORMONES IN INSECTS

The thoracotrophic hormone

The starting material for the extraction of thoracotrophic hormone should naturally be insect brains, and since hormones are active in minute quantities, very large numbers of brains must be used initially. The silk industry, based upon the silkmoth *Bombyx mori*, breeds enormous numbers of the insects every year, and it is not surprising that *Bombyx* has been the starting material for the extraction not only of thoracotrophic hormone but also of ecdysone (p. 218*ff*). Equally, the economic importance of *Bombyx* has led Japanese biologists to an intensive study of the insect.

Kobayashi and Kirimura in 1958 extracted thoracotrophic hormone from 8500 pupal brains of *Bombyx*, assaying their preparations upon brainless pupae—the positive result being the initiation of adult development. They eventually produced an oily material, soluble in organic solvents, and apparently lipid in nature. This result was surprising, since it had been expected that insect neurosecretory hormones, like those in the vertebrates, would prove to be polypeptides or low molecular weight proteins. But a few years later, 4 mg of the crystalline hormone was isolated from 220 000 *Bombyx* brains, dissected free from corpora cardiaca and allata.[174] This quantity of material was sufficient for its chemical characterization: it turned out to be identical with cholesterol (Fig. 12.1).

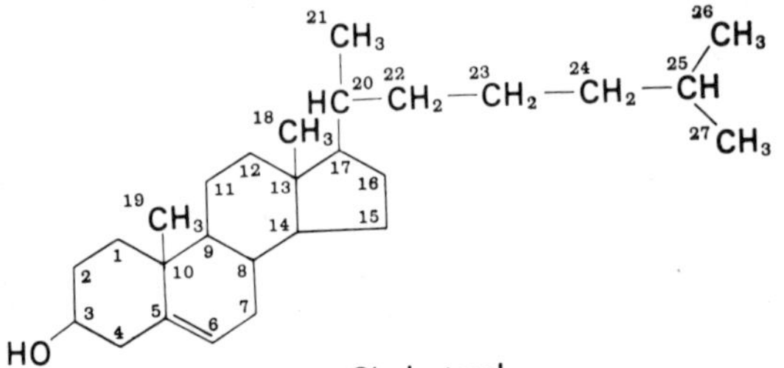

Fig. 12.1 The formula for cholesterol. The carbon atoms are numbered in the standard manner.

Since the thoracic gland hormone is a steroid (p. 219), it was thought significant that the thoracotrophic hormone could be the precursor for steroid biosynthesis.

When this identification of thoracotrophic hormone was published, many laboratories attempted to confirm it. Although in some instances very small quantities of commercial cholesterol would induce development in brainless insects, a normal dose-response relationship (increasing percentage development with increasing cholesterol concentration) could not be obtained[243] (Fig. 12.2). Very highly purified cholesterol often

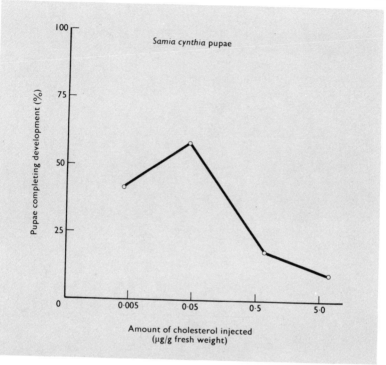

Fig. 12.2 Percentage development of *Samia* pupae with increasing dosage of cholesterol. Note the absence of a typical dose-response effect. Logarithmic scale on abscissa. (After Schneiderman and Gilbert[243])

produced no result. Moreover, many insects normally contain large quantities of cholesterol, so why the further addition of a minute amount of the substance should activate the thoracic glands was bewildering.

At about the same time that these conflicting results appeared, Ichikawa and Ishizaki[153] reported that they had extracted from *Bombyx* brains a water soluble hormone, stable within only a narrow range of pH, non-dialyzable, precipitated by ammonium sulphate and trichloroacetic acid, and which was inactivated by incubation with bacterial proteases. This hormone, therefore, has many of the properties of a protein.

It is difficult to reconcile the chemical natures of these two brain hormones from *Bombyx*. It is possible that the brainless pupae used for the bio-assay of the lipoidal extract were very sensitively poised on the brink of development. The addition of minute amounts of true brain hormone in the extracts could then initiate development; or perhaps contaminating steroids—present in both extracts and commercial cholesterol—activated the thoracic glands. Ecdysone itself will activate thoracic glands (Fig. 12.3). Certainly, a protein or polypeptide thoraco-

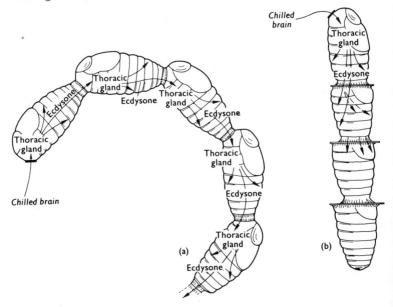

Fig. 12.3 (a) A chain of brainless diapausing pupae of *Hyalophora cecropia*. Implantation of a chilled brain (compare Fig. 8.1) into the first pupa terminates diapause in the whole parabiotic chain and adults develop from all pupae. (b) Chain of pupal abdomens of *Hyalophora* joined to a brainless pupa. When a chilled brain is implanted, only the first two members of the chain terminate diapause and develop.

It is concluded that in (a), ecdysone from the anterior pupa stimulates the thoracic glands of the next in line; ecdysone from this pupa stimulates the thoracic glands of the next, and so on. Each individual therefore has a full complement of the hormone. In (b), however, ecdysone from the single pair of thoracic glands is insufficient for the development of more than the first two individuals. (After Williams[287])

trophic hormone accords better with what is known of neurosecretory hormones in other animals. But until the hormone can be completely purified, its chemical composition known and the artificially synthesized compound shown to have the same biological effects as the natural product, the doubt must remain that perhaps a neurosecretory protein carrier is associated with a steroidal molecule. The difficulties inherent in purifying and characterizing protein hormones from vertebrates suggest that the problem of the chemical nature of the insect thoracotrophic hormone will take some time to solve.

Other insect neurosecretory hormones

Purified extracts of **bursicon** are stable up to temperatures of 55°C, but are inactivated by repeated freezing and thawing. High salt concentrations, extremes of pH and short-chain alcohols all inhibit bursicon activity. There is little doubt that bursicon is a protein hormone; its molecular weight is about 40 000.[207]

The **hyperglycaemic hormones** of cockroaches and locusts are soluble in methanol and water, can be separated chromatographically in solvent systems typically used for amino-acids and peptides, and are inactivated by the enzyme chymotrypsin. They are thus likely to be polypeptides. In the locust, the less potent hyperglycaemic hormone is present in the neurosecretory storage lobes of the corpora cardiaca and also in the pars intercerebralis of the brain: it is undoubtedly neurosecretory in origin. The more potent hormones can be extracted only from the glandular lobes of the corpora cardiaca: its origin is therefore not *immediately* neurosecretory, although it is likely that cerebral neurosecretion is concerned indirectly with its production. The situation recalls that of the glandular anterior pituitary producing polypeptide hormones under the control of neurosecretory releasing factors from the hypothalamus.

The locust **diuretic hormone** has the same physical properties as the neurosecretory hyperglycaemic hormone and is similarly inactivated by chymotrypsin. This hormone, too, is therefore likely to be a polypeptide. The hormones affecting the heart beat in both cockroaches and locusts, whether they control the heart directly or through the production of indolalkylamines by pericardial cells, also have many of the properties of polypeptides.

NEUROSECRETORY HORMONES IN CRUSTACEA

Presumably because of the ease of assaying chromatophorotrophic hormones in the crustaceans, the extraction, purification and identification of these hormones has proceeded more rapidly than that of the developmental hormones. The incubation of sinus glands of *Palaemon squilla* with extracts of hepatopancreas progressively inactivates the red pigment concentrating hormone. It is likely that this is due to enzymic digestion of the hormone. Pepsin, trypsin and acid hydrolysis also inactivate the hormone, suggesting the rupture of peptide bonds, and leading to the conclusion that the hormone is a polypeptide.[172] The melanin dispersing hormone of *Uca pugilator* is destroyed by chymotrypsin and hepatopancreas extracts.[221] In *Cambarellus*, electrophoresis of the red pigment dispersing hormone from the brain and circum-oesophageal connectives shows that the hormone is electropositive at pH 2·3 but electronegative at pH 7·7. This phenomenon of charge reversal at different pH strongly suggests that the hormone is a polypeptide. Such electrophoresis of eyestalk extracts at different pH values can also show that retinal pigment adapting hormones are distinct from the chromatophorotrophins.[90]

Evidence from both the insects and the crustaceans strongly supports the notion that the neurosecretory hormones are protein or polypeptide compounds. The arthropod neurosecretory mechanisms thus fall into line with those of the vertebrate groups, where neurosecretory hormones are of a similar chemical nature. A great deal of histochemical evidence exists for the presence of protein or lipo-protein substances within neurosecretory cells of many animals. This evidence has been deliberately disregarded because neurosecretory material is not necessarily related to the hormonally active products which are released into the blood. In fact, in the vertebrates the histologically and histochemically demonstrable neurophysine appears to be a carrier substance for the biologically active peptides and proteins. It is conceivable that in other animals, carrier neurosecretion could be associated instead with other active molecules. So far, the evidence in arthropods is generally against such an idea: but the possibility should always be borne in mind.

STEROID HORMONES IN ARTHROPODA

The insect moulting hormone: ecdysone

As early as 1935, a bio-assay for the estimation of moulting hormone extracts was devised. This consists of ligaturing blowfly larvae behind Weismann's ring before pupation: puparium formation proceeds normally anterior to the ligature, but is prevented posteriorly if the ligature is applied before hormone is released from the ring gland[94] (Fig. 12.4). Extracts containing moulting hormone injected into the abdomen of the larva will then induce pupation. With the exception that two ligatures are now applied, so that injections can be made through the second ligature into the abdomen without loss of fluid, the bio-assay has remained virtually unchanged for almost 40 years.

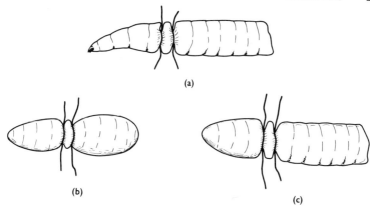

Fig. 12.4 Double ligature technique in the *Calliphora* larva bio-assay for ecdysone (a). If the ligatures were applied after endogenous ecdysone were released, both parts of the larva would pupate (b). If the ligatures are applied earlier, only the anterior part of the body pupates (c) and the posterior region can be injected with test solutions from a needle inserted between the two ligatures.

In 1954, Butenandt and Karlson extracted 25 mg of crystalline moulting hormone, which they called ecdysone, from 500 kg of *Bombyx* pupae.[33] The crystalline ecdysone induces pupation when as little as 0·0075 μg is injected into ligatured blowfly larvae. With most batches of larvae, 0·010 μg of ecdysone is effective; and this quantity of hormone is known as one ***Calliphora unit*** (C.U.).

Although ecdysone was isolated in crystalline form in 1954, the quantity extracted was too small for accurate chemical analysis with the means then available. But some 10 years later, starting with 1000 kg (dryweight) of Bombyx pupae (equivalent to about 4 tons fresh weight) 250 mg of ecdysone were prepared and the hormone was proved to be indisputably a steroid,[162] with the structure shown in Fig. 12.5. That one of the most

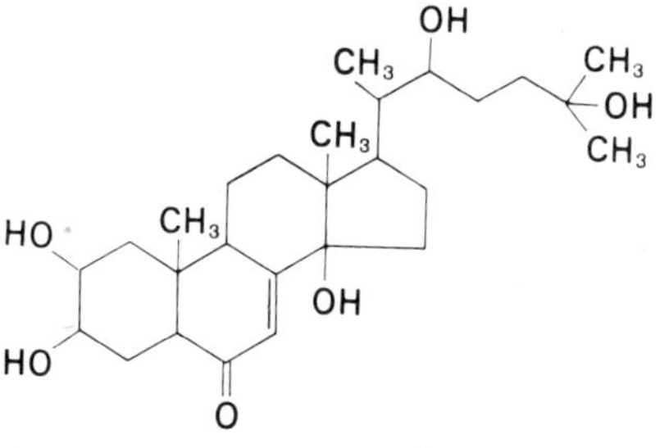

Fig. 12.5 Formula for α-ecdysone (compare with Fig. 12.1).

α-ecdysone

important of the insect developmental hormones is a member of a class of compounds well known and widespread in the vertebrate classes is of exceptional interest.

In fact, two ecdysones can be extracted from *Bombyx* pupae: that which has just been described is sometimes called α-ecdysone; a second steroid present in very much smaller quantity is called β-ecdysone. The same two ecdysones have now been isolated from the Moroccan locust, *Dociostaurus maroccanus*, and it is interesting that in this species β-ecdysone is the predominant steroid.[250] In the tobacco hornworm, *Manduca sexta*, α-ecdysone is also present in smaller quantity than a second ecdysone, which appears to be 20-hydroxyecdysone[159] (Fig. 12.6) and is

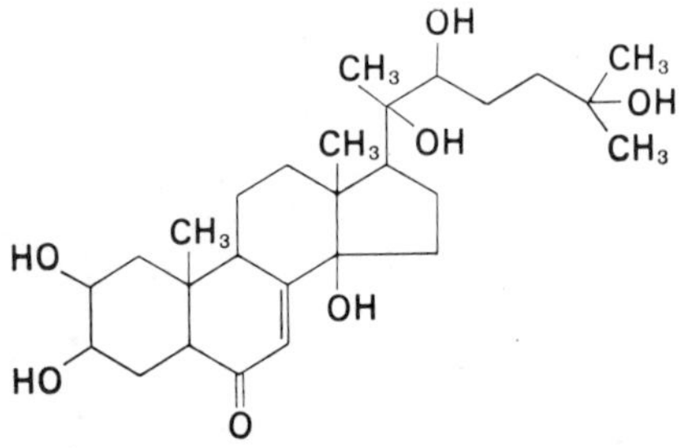

Fig. 12.6 The formula for 20-hydroxyecdysone. Note the additional -OH group on carbon atom 20.

20-hydroxyecdysone

probably identical with β-ecdysone. A third ecdysone, called ecdysterone, has been isolated from *Bombyx* extracts.[149] This has about five times the activity of α-ecdysone, although its structure seems to resemble that of 20-hydroxyecdysone: different stereoisomeric forms could explain such differences in activity.

Ecdysone has now been synthesized and is available in quantity, but in addition, large yields of ecdysones, with marked biological activities when tested on blowfly larvae, have been extracted from the leaves and roots of ferns such as *Podocarpus* and *Achyranthis*.[175, 150] Ponasterone A (Fig. 12.7) and inokosterone (Fig. 12.8) show similarity in structure to the insect's own ecdysones.

Ecdysone has the same carbon skeleton as cholesterol (Fig. 12.1). When radioactively labelled cholesterol is administered to *Calliphora* larvae, a great deal of the radioactivity is incorporated into ecdysone.[160, 161] Cholesterol is therefore very likely to be the precursor of ecdysone in insects. Insects cannot synthesize cholesterol from simpler precursors: however, phytophagous insects can convert plant sterols to cholesterol, and carnivorous insects will obtain cholesterol from their prey.

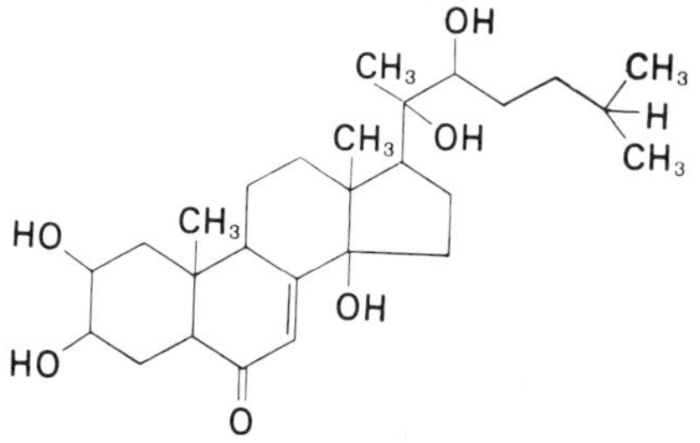

Fig. 12.7 The formula for Ponasterone A, a plant steroid with ecdysone activity.

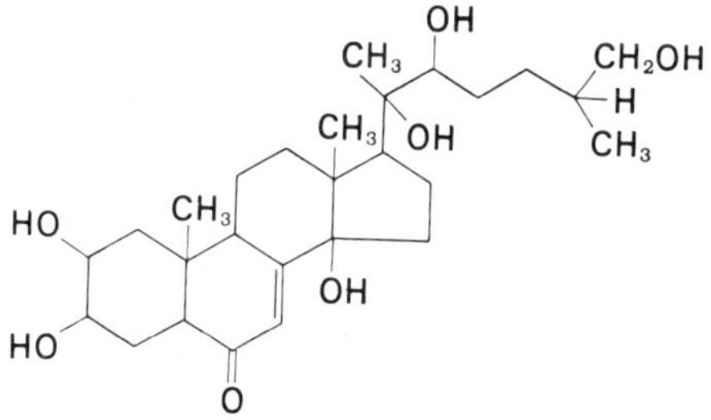

Ponasterone A

Fig. 12.8 The formula for inokosterone, a plant steroid with ecdysone activity.

Inokosterone

Crustacean moulting hormones

Ecdysone is able to accelerate moulting in some Crustacea, and crayfish homogenates, extracted for steroids, have some activity in the *Calliphora* larva test. These observations led to the idea of a crustacean steroid moulting hormone, which was called **crustecdysone**. This steroid has now been isolated and identified: it apparently has the same structure as β-ecdysone (20-hydroxyecdysone).[121] But in addition, a *second* ecdysone has been isolated from the marine crayfish *Jasus lalendei*, in which the hydroxyl group on C atom 3 is replaced by hydrogen: the compound is therefore called deoxycrustecdysone[101] (Fig. 12.9).

The existence of two or three (and perhaps many more) ecdysones in individual species of insects and crustaceans is of considerable interest. But at present it is not known whether the individual hormones have specific effects upon different aspects of metabolism and development, or whether they are in fact precursors or breakdown products of a single steroid hormone. Some water beetles are able to produce large quantities of steroids, which are identical with those produced by vertebrates, e.g. deoxycorticosterone[240] (Fig. 12.10). These are used in defence against fish predators. Thus steroid biosynthetic mechanisms in insects can be quite sophisticated. Following the initial major achievement of Karlson

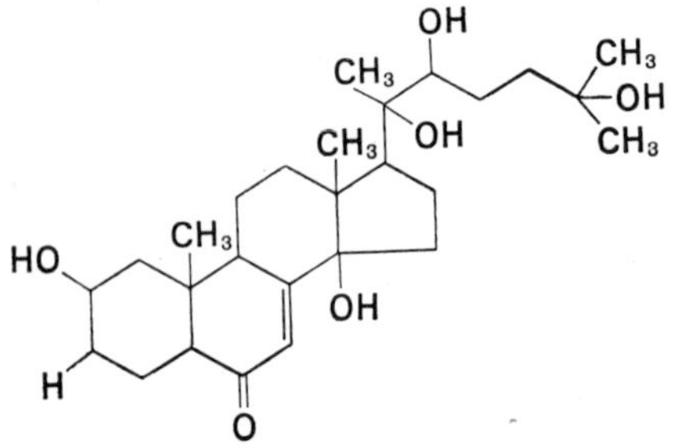

Fig. 12.9 The formula for deoxycrustecdysone, one of the moulting hormones of the marine crayfish.

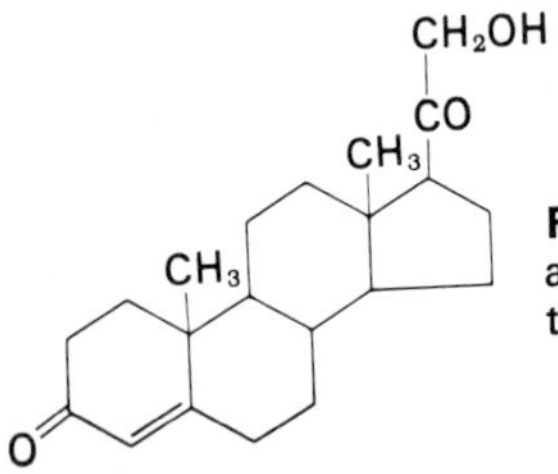

Deoxycrustecdysone

Fig. 12.10 The formula for deoxycorticosterone, a steroid produced by water beetles identical with that from the adrenal glands of vertebrates.

Deoxycorticosterone

and his colleagues in isolating and identifying ecdysone, studies on arthropod steroid hormones have already begun to increase and may soon seem as complicated as those in the vertebrate animals.

The crustacean androgenic gland hormone

Preliminary work suggested that the hormone from the androgenic gland was protein in nature. However, when the androgenic glands of *Ocypoda platyaris* are extracted for sterols, the non-saponifiable lipid fraction mimics the effects of androgenic gland transplants when injected into female crabs. Moreover, injected testosterone will convert ovarian to testis tissue, although it will not by itself induce the secondary male characters.[230]

In the lobster, *Homarus americanus*, testosterone is produced when androgenic glands are incubated with a precursor known to produce testosterone in vertebrate systems. The glands seem therefore to contain an enzyme capable of producing testosterone,[111] although this steroid has not yet been identified in the blood. The significance of mechanisms in invertebrate tissues which duplicate the enzymatic conversion found in vertebrates must always be viewed with caution until the biological function of the reactions are known. But the evidence so far suggests that the crustacean androgenic gland could produce a sex steroid, perhaps in combination with a protein hormone to fulfil the complete range of androgenic gland activities.

THE INSECT JUVENILE HORMONE

The first active extracts of juvenile hormone (neotenin) were made by Carroll Williams, from the abdomens of male *Hyalophora cecropia* moths where it accumulates in very large quantities (Fig. 6.11) (for an unknown reason).[289] A cold ether extract of a single moth abdomen yields about 200 mg of a yellow oil, which contains virtually all the lipids in the body, together with juvenile hormone. When 50 mg of the extract are injected into a pupa the individual moults into a second pupa—just as if active corpora allata had been implanted.[109] An even more sensitive bio-assay has been developed for both hemimetabolous and holometabolous insects. When extracts containing juvenile hormone activity are applied locally to the cuticle of last larval hemimetabolous or pupal holometabolous insects, the epidermal cells are induced to secrete another larval or pupal cuticle at the next moult, whereas the surrounding cells produce a normal adult cuticle[109] (Fig. 12.11).

Although very highly purified, the juvenile hormone from *Hyalophora cecropia* proved very difficult to characterize. In the meantime, a large number of compounds, and extracts from a multitude of organisms were assayed for juvenile hormone activity with a surprisingly large number of positive results[244] (Fig. 12.12). Of particular interest was the discovery in 1959 by Karlson and Schmialek that extracts of insect faeces often possessed considerable juvenile hormone activity.[164] In 1961 Schmialek extracted 80 kg of the faeces of the mealworm, *Tenebrio molitor*, and identified the active agent as the terpene farnesol and its oxidation product farnesal[241] (Fig. 12.13a). In *Tenebrio, Rhodnius*, the cockroach and several other insects, farnesol and particularly its methyl ether and its diethylamine derivative (Fig. 12.13a), are able to reproduce completely all the effects of juvenile hormone: delayed metamorphosis in *Rhodnius* and reversal of adult characters in a moulting adult[281]; gonadotrophic activity in several insects; prolongation of diapause in insects such as the rice stem borer, where the corpora allata are important in its control[96]; and in many insects farnesol replicated the cuticle bio-assay in a manner exactly similar to extracts of juvenile hormone. It seemed conclusive that the insect juvenile hormone was farnesol, or some very closely related terpenoid compound. When Schmialek[242] isolated

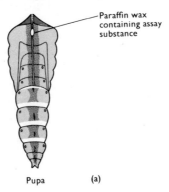

Paraffin wax containing assay substance

Pupa   (a)

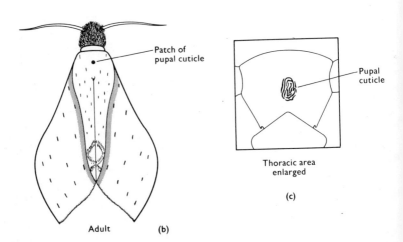

Patch of pupal cuticle

Pupal cuticle

Thoracic area enlarged

(c)

Adult   (b)

Fig. 12.11 The *Galleria* wax:cuticle test for juvenile hormone activity. The compound to be assayed is dissolved in glycerol contained in a small disc of paraffin wax. A small area of cuticle and epidermis is removed from a *Galleria* pupa, and the wound sealed with the melted disc of wax (a). The area beneath the wax produces another pupal cuticle during the pupa-adult transformation if the assay compound has juvenile hormone activity (b), (c) shows such an area at greater magnification.

Organisms, extracts of which have juvenile hormone activity	Compounds with juvenile hormone activity
Ochromonas malhamensis (Protozoa) *Tetrahymena pyriformis* (Protozoa) Brewer's yeast (*Saccharomyces*) *Tubularia crocea* (Coelenterata) *Physalia physalis* (Coelenterata) *Chaetopterus variopedatus* (Annelida) *Nereis virens* (Annelida) *Venus merchraria* (Mollusca) *Uca pugilator* (Crustacea) *Homarus americanus* (Crustacea)	Phytol, nerolidol, dodecyl methyl ether, dodecyl ethyl ether, dodecyl methyl sulphoxide, decyl methyl sulphoxide, 10-undecylenic methyl ether, farnesol, farnesyl methyl ether, farnesyl diethylamine, farnesal

Fig. 12.12   Extracts of organisms and chemical compounds which have juvenile hormone activity. (After Schneiderman *et al.*[244])

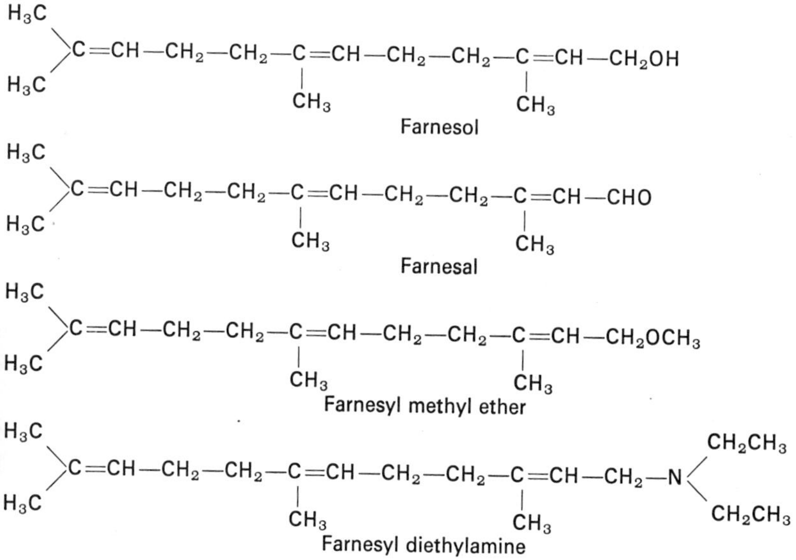

Farnesol

Farnesal

Farnesyl methyl ether

Farnesyl diethylamine

Fig. 12.13(a)   The structures of farnesol, farnesal, farnesyl methyl ether and farnesyl diethylamine: compounds with high juvenile hormone activity.

farnesol from extracts of male *Hyalophora cecropia* in 1963 the issue seemed to be beyond doubt.

But farnesol and its derivatives are many times less effective than highly purified extracts from *Hyalophora cecropia*[243, 107] (Fig. 12.13b).

Substance	Juvenile hormone activity (*Galleria* units/g)
Farnesol	16
Farnesal	32
Farnesyl methyl ether	1000–1500
Farnesyl diethylamine	1000–1500
Crude *H. cecropia* extract	1000
Concentrated *H. cecropia* extract	300 000 000

Fig. 12.13(b) The relative juvenile hormone activity of farnesol and its derivatives compared with extracts from adult males of *Hyalophora*. (From Gilbert and Goodfellow[107])

Moreover, farnesol is present in larger amounts in chilled pupae of *Hyalophora* than in adult males, although the pupae possess no extractable juvenile hormone, while adult males contain enormous amounts of juvenile hormone (Fig. 6.11). Finally, removal of the corpora allata from *Hyalophora* pupae prevents completely the appearance of juvenile hormone in the adult males, but their content of farnesol is unaltered.[117, 107] These results contradict completely the idea that farnesol actually is the juvenile hormone. Farnesol and its derivatives *mimic* the effects of juvenile hormone, more so in some insects than in others, but they are not themselves the true juvenile hormone.

In 1967, Röller and his colleagues in the University of Wisconsin published the definitive structure of insect juvenile hormone.[228] It is methyl 10,11-epoxy-7-ethyl-3,11-dimethyl-2,6-tridecadienoate (Fig. 12.14).

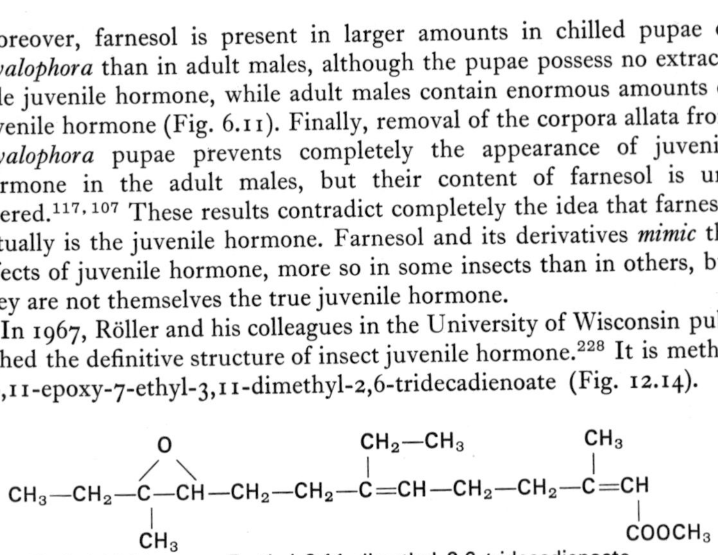

Methyl 10,11-epoxy-7-ethyl-3,11-dimethyl-2,6-tridecadienoate

Fig. 12.14 The structure of juvenile hormone as given by Roller *et al.*[228]

With two double bonds and its cyclic oxirane component, the molecule can exist in 16 different stereoisomeric forms, only one of which can be the authentic hormone. Moreover, with ethyl groups on carbons 7 and 11 (Fig. 12.14) the compound cannot be synthesized from any known terpenoid, although similar molecules with more or less juvenile hormone activity have been produced[65] (Fig. 12.15).

During normal insect development, juvenile hormone is necessary during precisely determined periods (see p. 84); its continuing presence, for example, during the last larval instar, can prevent metamorphosis and

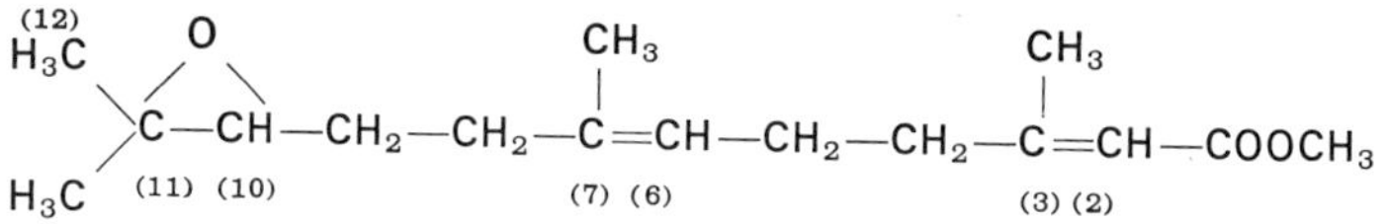

Methyl 10-epoxy-3,7,11-trimethyldodeca-2,6-dienoate

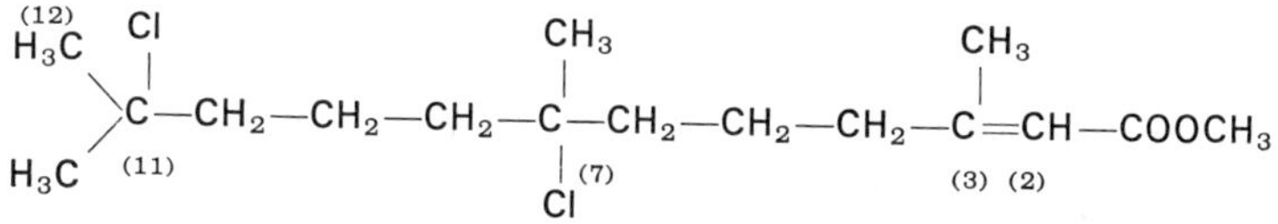

Methyl 7,11 dichloro-3,7,11-trimethyldodeca-2-enoate

Fig. 12.15 The structures of two highly active juvenile hormone mimics.

lead to the death of the insect without its becoming adult. In particular, juvenile hormone must be absent from the eggs in many species or embryonic development is abnormal and the larvae do not hatch.

It has long been envisaged that juvenile hormone, and its mimics, could be used as insecticides by administering them at the wrong times in insect life-histories. Such compounds would have a tremendous advantage over insecticides at present used, since the insects would be unlikely to develop immunity to substances which play a natural role in their own lives. Some juvenile hormone mimics are more effective in some species than in others, so that the development of *specific* insect pesticides could also be possible—in contrast to compounds like DDT which kill good and bad insects alike.

The feasibility of such developments was confirmed in a startling way when Dr. Karel Slama of the Charles University in Prague spent a year working at Harvard with Professor Carroll Williams. Slama took a stock of *Pyrrhocoris apterus* with him, but found that in Harvard the bugs would not develop into normal adults. Instead, at the end of the 5th larval instar, the insects all moulted into a supernumerary 6th larval instar, or into adultoids which preserved many larval characters. A few even grew and moulted into 7th instar larvae (Fig. 12.16). By analogy with the effects of implanting corpora allata into last instar larvae (see Chapter 6), it seemed likely that the *Pyrrhocoris* were being unintentionally exposed to a source of juvenile hormone, or a substance with juvenile hormone activity.

A systematic comparison of the culture conditions in Harvard with those in Prague eventually revealed that paper towels in the rearing jars were the source of the substance.[248, 249] The chemical was not added during manufacture because all papers, including newspapers, of American origin had the same effect upon *Pyrrhocoris*. Papers of

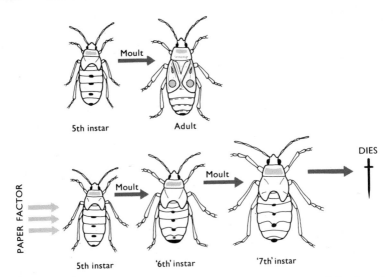

Fig. 12.16 The action of paper factor upon the development of *Pyrrhocoris apterus*.
Upper row : normal moult from last instar larva to adult.
Lower row : paper factor induces the development of two giant supernumerary larvae followed by death during the adult moult. (After Slama and Williams[249])

European or Japanese manufacture had no effect. The major source of wood pulp for paper manufacture in the United States is the balsam fir (*Abies balsamea*), and extracts from this tree proved to contain high juvenile hormone activity when assayed on *Pyrrhocoris*. Hemlock and yew also showed high activity, but red spruce, American larch and southern pine had little activity. The substance was called **paper factor** by Slama and Williams; it has now been identified (Fig. 12.17). It shows some resemblance to juvenile hormone.

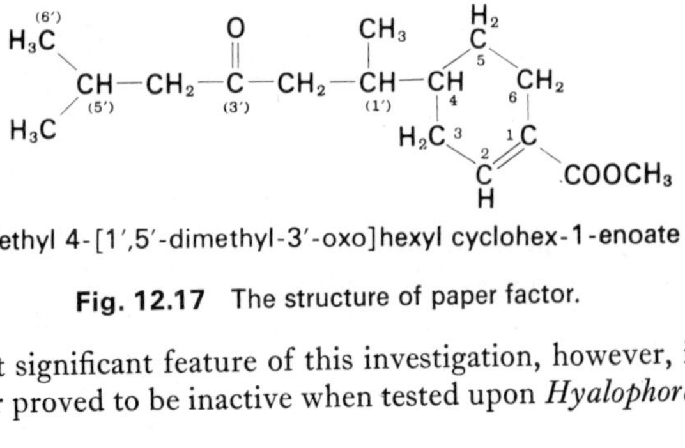

Methyl 4-[1′,5′-dimethyl-3′-oxo]hexyl cyclohex-1-enoate

Fig. 12.17 The structure of paper factor.

The most significant feature of this investigation, however, is that the paper factor proved to be inactive when tested upon *Hyalophora cecropia*,

Telea polyphemus, and even upon *Oncopeltus fasciatus* and *Rhodnius prolixus*—bugs belonging to the same Order as *Pyrrhocoris*. This immediately raises the possibility of using such chemicals as specific insect pesticides. Slama and Williams point out that the family Pyrrhocoridae include a number of pests, such as *Dysdercus cingulatus*, and other cotton stainers, which can ruin cotton crops.

Such insecticides can have two general applications: control against pests of stored products, and against those of field crops and animals. Laboratory experiments in which a juvenile hormone analogue is mixed with grain have already shown that infestations of *Tribolium castaneum* and other insect pests can be reduced or eliminated.[259] This method of control is likely to be widely successful for stored products, where usually *all* the insects present are undesirable. But in addition a technique has been developed which could provide a quite specific control of insect pests in the field.[202]

The method is effectively an extension of the sterile male technique so successfully used against the screw-worm, *Cochliomyia americana*, in America. This involves rearing large numbers of flies, irradiating the males to sterilize them, and releasing them in large numbers at the appropriate time. If an area is saturated with sterile males, the females will mate with them and lay infertile eggs. The female screw-worm mates only once: clearly this method of control is much less effective in insect species in which the female mates many times. The population of sterile males would have to be kept up for such extended periods that the numbers involved would become practically and economically impossible.

But in *Pyrrhocoris apterus*, for example, a juvenile hormone analogue such as the dihydrochloride of methyl farnesoate (DMF) seriously disturbs embryonic development when applied to the eggs, and less than 1 μg applied to a mature female causes her to lay inviable eggs for the rest of her life. The males of *Pyrrhocoris* will tolerate up to 1000 μg of DMF applied topically—and they will pass sufficient DMF to females during mating to sterilize them for life. Males with as little as 100 μg each of DMF will pass 1–5 μg to females when kept with them for only 4 hours. With the much more effective compounds now being developed, even 10 μg per male will be sufficient.[202]

This method has the great advantage that a female once mated with a treated male will be sterile *for life*, no matter if she subsequently mates with any number of untreated, normal males: it extends the exact specificity of the screw-worm kind of pest control to species with polygamic females.

'Hormonal' insecticides, with all their advantages of specific control over individual pests, coupled with their presumed ability not to induce immunity reactions, are likely to be the most effective ever developed. In terms of increased production and reduced wastage of food, their value will be incalculable.

13

Pheromones

Pheromones are defined as substances produced by one individual of a species which cause particular behavioural and/or developmental reactions when received by another individual of the same species.[163] They are effective in minute amounts and act as chemical messengers *between* individuals in much the same way as hormones *within* an individual; in fact, they are sometimes called ectohormones. This account will deal mainly with those pheromones in which some interaction with the endocrine system has been demonstrated. The alarm substances of ants and earthworms, trail markers, territorial chemicals, and many other pheromones concerned predominantly with behavioural reactions will be disregarded. Even so, developmental pheromones often have a large behavioural component. Moreover, only in the insects among the invertebrates have both hormones and pheromones been studied in detail, so this group inevitably predominates.

Sex attractants in insects

Many insects produce chemicals to attract from a distance members of the opposite sex. The first such attractant to be identified was that of the female silkmoth, *Bombyx mori*, consequently called **bombykol** (Fig. 13.1). **Gyptol** from the female gypsy moth, *Porthetria dispar*, has also been identified[155] (Fig. 13.1); a similar compound has been synthesized (**gyplure**: Fig. 13.1), with two additional CH_2 groups after the double bond, which is even more attractive to male gypsy moths than the natural compound. Males will respond to as little as 10^{-12} µg of bombykol or

$$CH_3—CH_2—CH_2—CH=CH—CH=CH—(CH_2)_8—CH_2OH$$

Bombykol

$$CH_3—(CH_2)_5—CH—CH_2—CH=CH—(CH_2)_5—CH_2OH$$
$$\mid$$
$$COOCH_3$$

Gyptol

$$CH_3—(CH_2)_5—CH—CH_2—CH=CH—(CH_2)_7—CH_2OH$$
$$\mid$$
$$COOCH_3$$

Gyplure

Fig. 13.1 Structures of bombykol and gyptol, the sex attractants of the silkmoth and gypsy moth, together with the synthetic compound gyplure which has two more -CH_2- units than the natural compound.

gyptol. The gypsy moth is a serious pest of both evergreen and deciduous trees, its caterpillar stripping the foliage and often causing the death of the trees. Its control by luring and trapping large numbers of males with gyptol and gyplure is already being investigated: such a method of control of this and other insect pests is likely to have an important future.

Female cockroaches produce sex attractants from one or more anal, sternal and tergal glands. In the Cuban cockroach, *Byrsotria fumigata*, removal of the corpora allata from the female shortly after the imaginal moult prevents the production of sex pheromones, and the allatectomized females remain unattractive to the males and do not elicit the masculine precopulatory behaviour. Reimplantation of corpora allata is followed by pheromone production within 10–16 days.[6] Moreover, in *Periplaneta americana* the production of sex pheromone ceases during the time the ootheca is being formed and extruded—when mating is impossible. In these cockroaches, oocyte development is controlled by the corpora allata, so that the concomitant endocrine control of pheromone production provides a mating signal during the correct period. Support for this idea comes from experiments with the cockroach *Pycnoscelus surinamensis*, which has bisexual and parthenogenetic strains. Allatectomy of females of bisexual *Pycnoscelus* prevents the production of sex pheromones as it does in *Byrsotria*; but allatectomy of the parthenogenetic females has no effect upon sex pheromone production. Further, females of the bisexual strain do not produce pheromone when they are carrying an egg case: those of the parthenogenetic strain do. Oocyte development is controlled by the corpora allata in both bisexual and parthenogenetic *Pycnoscelus*; but when, as in the parthenogenetic females, a mating signal is no longer required, the endocrine co-ordination of mating with oocyte development is lost.[7]

In *Bombyx mori*, *Antheraea pernyi* (the oak silkworm), *Galleria mellonella* (the waxmoth) and other lepidopterans with short-lived adult stages, the adult endocrine system controls neither sex pheromone production nor oocyte development. So it is likely that only those insects which have regularly repeated reproductive cycles, with intermittent periods when mating is possible, will possess an endocrine mechanism co-ordinating the two events.

In many Orthoptera, either the male or the female (and sometimes both) produces a mating song by rubbing one part of the body against another, frequently the legs against the elytra, or the elytra against each other. This stridulatory song is an auditory communication mechanism comparable with the olfactory ones described above. The corpus allatum hormone co-ordinates the onset and termination of female stridulation with oocyte development.[193] Although affecting central nervous activity in the one instance, and glandular biosynthesis in the other, a signal is sent out at the appropriate time. But chemical signals can often acquire greater versatility than communication by sound. In the honey bee, *Apis mellifera*, the drones are attracted by a queen when she is flying in the afternoon of a warm, sunny day and only when she is at a certain height relative to the wind-velocity.[34] The queen's sex pheromone is only stimulatory when other sensory impulses are also passing to the drones' central nervous systems. Drones are not attracted to a queen in the hive or on the alighting board. This could be a device to prevent colony inbreeding. The honey bee's sex attractant is 9-oxodecenoic acid[34] (Fig. 13.2): but in addition to its sex pheromone activity, stimulating the

$$CH_3—\overset{\overset{\displaystyle O}{\|}}{C}—(CH_2)_5—CH=CH—COOH$$
9-oxodecenoic acid

$$CH_3—\overset{\overset{\displaystyle OH}{|}}{CH}—(CH_2)_5—CH=CH—COOH$$
9-hydroxydecenoic acid

Fig. 13.2   9-oxodecenoic and 9-hydroxydecenoic acids: honeybee pheromones.

drone's olfactory receptors, the same substance plays important developmental and behavioural roles *within* the hive, primarily affecting the (female) worker bees.

Maturation pheromones

The queen is the only reproductive female in the honey bee hive. In the remaining females—the workers—oogenesis is inhibited. When the queen is removed from the hive, two important things happen: the workers modify one or more of the worker cells into the larger queen cell, and the ovaries of many of the workers begin to develop. There is thus both a behavioural and a developmental response by the workers to the disappearance of their queen.

The inhibition of both oogenesis and queen rearing in workers is normally caused by queen substance produced in the queen's mandibular glands, spread over the body during grooming, and licked off by attendant workers who share it with their colleagues in regurgitated food.[34] Queen substance is largely composed of 9-oxodecenoic acid, but this substance alone does not completely inhibit oogenesis. It has to be combined with inhibitory scent, which has very recently been identified as the odour of 9-hydroxydecenoic acid[35] (Fig. 13.2).

How these substances act in inhibiting oogenesis is unknown. 9-oxodecenoic acid can be identified in the crops of worker bees. But whether it stimulates specific chemoreceptors during feeding, thereby affecting the endocrine system, or whether it inhibits oogenesis directly after passing into the blood is conjectural. The behavioural response to the two substances is more likely to be due to specific sensory stimulation affecting central nervous activity.

9-hydroxydecenoic acid and 9-oxodecenoic acid together in a queenless hive are not quite as effective in inhibiting queen-rearing as a live mated queen. They are less attractive than a queen, and fewer individuals are consequently influenced by the substances. The queen, *in the hive*, presumably secretes another attractive substance.[34] But the two acids together enable *swarming* bees to find and remain with their queen.[34]

Old queens become deficient in the two acids and are superseded by new queens, when the workers react to the decreasing concentrations of the substances. Similarly, when the hive becomes overcrowded, workers may obtain less queen substance and swarm. In *Apis mellifera*, the versatility of 9-oxodecenoic acid, alone or in combination with other compounds, in influencing different kinds of behaviour, and development, must be acknowledged.

In the dry wood termite, *Kalotermes flavicollis*, there is a pair of royal reproductives, the king and the queen. The worker caste is actually composed of juvenile forms, consequently called pseudoworkers or pseudergates. These can develop into soldiers, or in the absence of the royal pair into secondary reproductive forms (Fig. 13.3).

In a normal colony of *Kalotermes*, the development of secondary reproductives from pseudergates is inhibited by a pheromone present in the excrement of the queen and distributed among the workers.[196] For the complete inhibition of secondary reproductives, the king must also be present. It seems that each member of the royal pair inhibits the development of reproductives of its own sex, and that each stimulates the production of pheromone by the other (Fig. 13.4). Moreover, the king produces another pheromone which stimulates the production of female reproductives in the absence of the queen.[196] Other pheromones determine the soldier caste, and still more are involved in the elimination of excess numbers of any one caste.

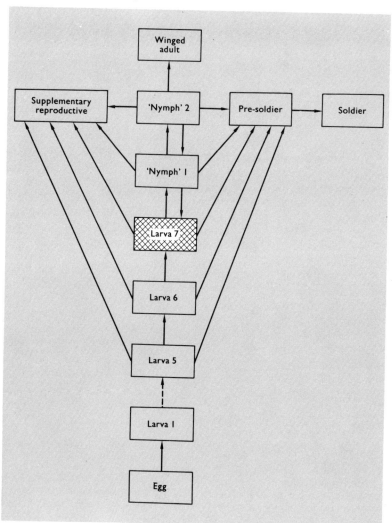

Fig. 13.3 Diagram of the caste structure of the termite *Kalotermes flavicollis*. The late larvae are 'false workers' or pseudergates. Note that regressive moults from nymphs to larvae can occur.

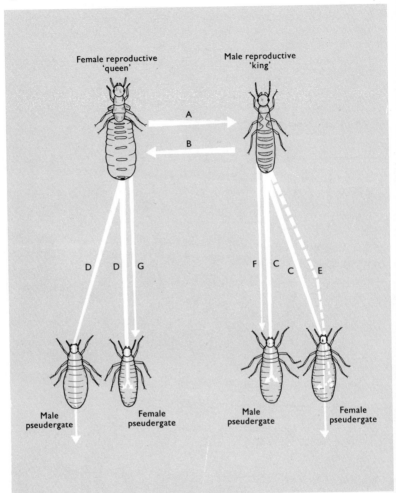

Fig. 13.4. Pheromones and the development of supplementary reproductives in *Kalotermes flavicollis*. The male reproductive ('king') produces a pheromone (C) which prevents reproductive development in male pseudergates but is without effect upon the female pseudergate. The female reproductive ('queen') similarly produces a pheromone (D) which inhibits reproductive development in female pseudergates, leaving the males unaffected. In the absence of the queen, the king produces a pheromone (E) which stimulates reproductive development in female pseudergates. The royal pair are reciprocally stimulated to produce their pheromones C and D by other pheromones (A and B). Yet other pheromones (F and G) affect the *behaviour* of pseudergates to eliminate an excess of supplementary reproductives if these develop. (After Luscher[196])

Although the evidence so far is slight, it is very likely that the pheromones work through the endocrine systems of the recipients. For example, the development of pseudergates to first and second stage nymphs is reversible, and hormones must be involved in such morphological plasticity. Soldiers can be developed from pseudergates by implanting the corpora allata of reproductives.[196] Moreover, competent pseudergates moult very soon after the removal of the royal pair, implying a rather direct effect upon the brain—thoracic gland system.[196] *Kalotermes flavicollis* forms small colonies with a limited number of castes. The pheromone-hormone interactions in species with large colonies and many more castes must be extremely complicated.

In the desert locust, *Schistocerca gregaria*, mature males produce a pheromone which accelerates maturation in other males and in females. The pheromone is not produced by allatectomized males, but the subsequent reimplantation of corpora allata reinitiates pheromone production[192] (Fig. 13.5). Maturation itself is controlled by the corpora allata and the cerebral neurosecretory systems (see p. 113) so the maturation

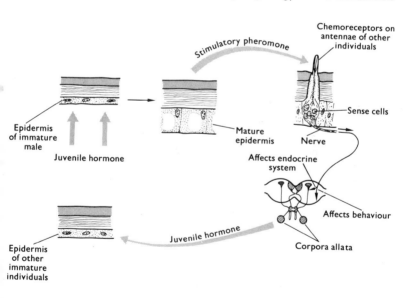

Fig. 13.5 The maturation pheromone of the desert locust, *Schistocerca gregaria*. Under the influence of juvenile hormone, the epidermis of one individual produces pheromone which stimulates the endocrine systems of other individuals probably by way of chemoreceptors and the nervous system. The consequent production of juvenile hormone in the recipients stimulates the production of more pheromone (see Loher[192]).

pheromone in some way stimulates the activity of the neuroendocrine system. It is not known whether this activation is direct, the pheromone passing through the recipient's cuticle into the blood, or indirect by stimulating specific chemoreceptors whose impulses eventually affect the neurosecretory system. The pheromone can elicit an immediate behavioural response in the recipients (twitching of the palps of the mouthparts and a kicking of the hindlegs[192]), so the central nervous system is certainly involved in at least part of the effect of the pheromone.

The net result of the stimulatory pheromone in the desert locust is that a rapidly maturing individual hastens the development of the more slowly maturing members of a group. In the African migratory locust, *Locusta migratoria*, it seems probable that an inhibitory pheromone is produced by immature adults which suppresses the potentially more rapid development of some individuals. But in both *Schistocerca* and *Locusta* the presence of the pheromones tends to synchronize reproductive development in all members of a group, although acting in quite opposite ways.

There are many obvious differences in morphology, physiology and behaviour between solitary and gregarious (swarming) locusts. Many of these differences can be traced to variation in the activities of parts of the endocrine system. The possibility that gregarization is caused by a specific pheromone will perhaps provide the best illustration of the relationships between endocrine glands and pheromones in insects.

Pheromones form only part of the many environmental influences which can affect development in insects, although in some instances their role is a major one. In social insects, where the colony may be considered a superorganism, the number and variety of pheromones may rival those of hormones in a highly developed animal. In many respects, the colony tends to isolate itself from the external environment, making contact at well defined points. The increased importance of pheromones in these situations is not surprising. Although it may not be wise to draw too close a parallel between the action of social pheromones upon the individuals of the colonial superorganism and that of hormones upon the cells and tissues of an individual, such a comparison does accentuate some common features of chemical communication systems in animals.

14

Hormones and the Environment

The ubiquity of neurosecretory mechanisms in invertebrates is confirmed by the results of many experiments upon the endocrine control of developmental and physiological processes. In the coelenterates and annelids, neurosecretory hormones appear to be the only endocrine co-ordinating principles present. The same is probably true of the flat-worms, nemerteans and nematodes although much less detailed information is available. In the molluscs generally, neurosecretions must also be of considerable importance but at present experimental evidence for their functions is sparse. The highly developed cephalopod molluscs possess epithelial endocrine glands, as do the equally complicated arthro-pods. It seems likely that epithelial gland hormones become necessary once a particular grade of structural organization is achieved. The vertebrate animals, well above this level of organizational complexity, have the most highly developed system of epithelial endocrine glands.

In the less highly organized invertebrates, neurosecretory hormones control directly the processes of growth, reproduction, water balance and so on. In the arthropods, too, neurosecretions often have a direct effect upon particular targets—for example, the chromatophorotrophins of crustaceans, the diuretic hormones of insects, and neurosecretory hor-mones affecting synthetic mechanisms in the insect fatbody and the crustacean hepatopancreas. But particularly in their control of develop-ment, moulting and reproduction, arthropod neurosecretory hormones act through epithelial endocrine glands. In insects, ecdysone is secreted by the thoracic gland, the activity of which is stimulated by the neuro-secretory thoracotrophic hormone; in crustaceans, crustecdysone secretion by the Y-organs is inhibited by the neurosecretory moult-inhibiting hormone. The insects' corpora allata are affected directly or indirectly by neurosecretory allatotrophic hormones: in the crustaceans, it is possible that metamorphosis is controlled directly by sinus gland hormones other than those which affect moulting in the post-meta-morphic stages. But the endocrine activities of the crustacean ovaries and androgenic glands are again under neurosecretory control.

In the less highly organized invertebrates, therefore, neurosecretory hormones act directly to control growth and development, and other shorter-term processes too. In the arthropods, although some events are still controlled directly by neurosecretory hormones, others, particularly growth, moulting and differentiation, are controlled indirectly by these hormones, with epithelial endocrine glands intervening in the system. In the vertebrate animals, the neurosecretory hormones like oxytocin and vasopressin have *direct* effects upon physiological mechanisms, but hypothalamic neurosecretory hormones exert trophic effects upon the anterior pituitary whose own hormones control the activities of other endocrine glands.

There is thus one step neurosecretory control in some invertebrates, two step control in more highly organized invertebrates, and three step control in the vertebrate animals [237] (Fig. 14.1). But direct neurosecretory control over *some* processes is still exercised even in the arthropods and vertebrates. The cephalopod molluscs, of the animals so far investigated, seem to be unique in that neurosecretory mechanisms play no part in the control of development. In the insects, sustained nervous inhibition of the corpora allata is thought to be unlikely: whether nervous inhibition of the cephalopod optic glands is more likely must remain debatable.

There are, of course, many endocrine mechanisms which are without any neurosecretory control. Hormone production by the gut and its associated glands in vertebrate animals is a good example of such inde-pendence. But in general, the endocrine control of growth and develop-ment in most animals seems ultimately to involve neurosecretory hormones. It has already been postulated that neurosecretory cells are the oldest hormone-producing cells in the animal kingdom, and that they function alongside purely nervous elements to co-ordinate responses in the individual with environmental change (Chapter 1). What is the evidence that neurosecretion functions in this way in modern animals?

Arguments in favour of the general hypothesis that neurosecretory mechanisms transform environmental stimuli to hormonal messages are often incomplete. Where a developmental process varies with the en-vironmental situation and can be shown to be under neurosecretory control, it is often assumed that the neurosecretory mechanism is influenced by the environment. But in the vertebrates, reciprocal inter-

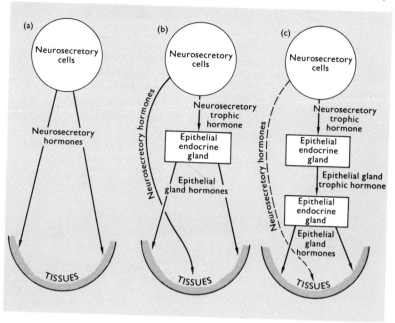

Fig. 14.1 (a) One step neurosecretory mechanism where neurosecretory hormones control developmental and physiological processes directly (e.g., coelenterates, annelids). (b) Two step neurosecretory mechanism where neurosecretory hormones exert *trophic* effects upon epithelial endocrine glands, hormones from which subsequently affect development and physiology. Note that elements of the one step system still occur (e.g., insects, crustaceans). (c) Three step neurosecretory mechanism, in which neurosecretory hormones exert trophic effects upon epithelial endocrine glands the hormones from which are themselves *trophic* to control the activity of yet further epithelial endocrine glands. Elements of the one and two step systems still occur (e.g., vertebrates). (See Scharrer and Scharrer[237] for a full discussion.)

actions between epithelial endocrine glands and hypothalamic neurosecretion are well established. In the insects, too, reciprocity between the corpora allata and cerebral neurosecretion is very likely (p. 122). Consequently, environmental effects upon developmental processes, such as the influence of temperature upon gonadal development in amphibians or insects, could *subsequently* cause changes in neurosecretory activity.

But there exist several very good examples of an immediate control over neurosecretory mechanisms by environmental stimuli. The best documented is perhaps the effect of photoperiod and temperature on the induction and termination of pupal diapause in insects (p. 143). There is now little doubt that the cerebral neurosecretory cells react to these environmental features and are the prime movers in the events that characterize diapause. The influence of direct and reflected light in controlling the release of chromatophorotrophins in crustaceans also indicates a fairly direct relationship between an environmental stimulus and neurosecretory activity (Fig. 11.13). The mammalian suckling reflex illustrates very well how an exogenous stimulus releases a neurosecretory hormone (oxytocin) which has a very rapid effect upon milk ejection from the mammary glands, but also a longer term effect in the continued secretion of prolactin for milk production (Fig. 14.2). Prolactin is an anterior pituitary hormone whose production and release is controlled by a neurosecretory hormone in the median eminence (Fig. 14.2).

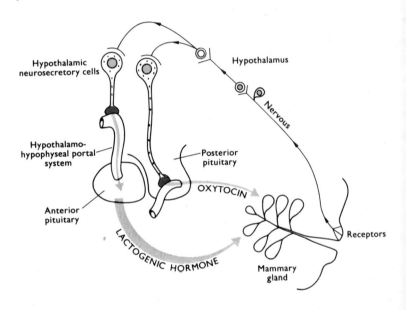

Fig. 14.2 The suckling reflex in a mammal. Both anterior and posterior pituitary hormones are secreted in response to stimulation of the nipple. Oxytocin release is very rapid and causes myoepithelial cells of the mammary alveoli to contract and force out the contained milk. Lactogenic hormone (among others) production and release is maintained by suckling stimuli so that milk synthesis is continued.

In *Rhodnius*, nervous impulses from an abdomen stretched with food cause the rapid release of thoracotrophic hormone. In the desert locust, low frequency electrical stimulation of the central nervous system for a short time accelerates the movement of material within the cerebral neurosecretory system[132] (Plate 3). Continued low frequency stimulation,

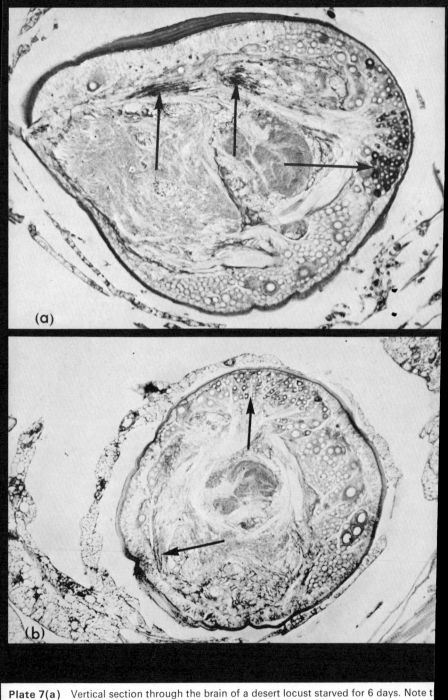

Plate 7(a) Vertical section through the brain of a desert locust starved for 6 days. Note t accumulation of material in many of the neurosecretory cells and their axons (arrowed).

Plate 7(b) Vertical section through the brain of a desert locust starved for 6 days and su sequently fed for 4 hours. Neurosecretion is present in much smaller amounts within t cells and their axons (arrowed) compared with those in Plate 7a. Neurosecretion disappe similarly after copulation, oviposition, etc.

or high frequency stimulation for a shorter time, releases the material from the system[133] (Plate 4). Stimulation of sensory receptors by tumbling the locusts in a jar has an effect upon the neurosecretory system similar to that of electrical stimulation. These techniques mimic natural sensory stimulation, because copulation in female locusts previously kept without males,[141] oviposition,[133] feeding after previous starvation,[140, 148] and flying,[138] all bring about the rapid release of material from the neuro-secretory system (Plate 7). In the vertebrates too, electrical stimulation of the hypothalamus releases both posterior and anterior pituitary hormones. That neurosecretory cells react to incoming nervous impulses by releasing hormones is beyond doubt. But *how* nervous activity accelerates the release of the elementary neurosecretory particles is unknown. Ultra-structural observations suggest that the particle of material is released from its surrounding membrane and passes through the axon terminal in a dispersed form (Fig. 14.3). But the relationship between this process and electrical activity in the axon is obscure.

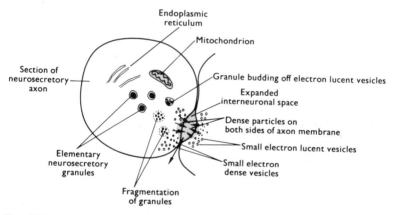

Fig. 14.3 Ultrastructure of a release site in a neurosecretory axon. The elementary neurosecretory granules appear to fragment and pass from the axon to the interneuronal space in diffuse form. (After Scharrer[236])

A variety of environmental factors affects neurosecretory activity in insects (Fig. 14.4); some, such as temperature and perhaps photoperiod, act directly upon the neurosecretory cell, others stimulate sensory receptors to control the cell through electrical activity within the central nervous system. The situation is similar in the crustaceans. In vertebrate animals, also, the environmental control of neurosecretory activity is well founded upon experiment and observation. In the less highly organized invertebrates, evidence for the exogenous control of neurosecretory

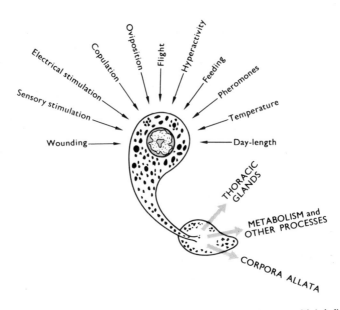

Fig. 14.4 Some of a variety of environmental and other factors which influence the release and/or synthesis of neurosecretions in insects. (After Highnam[136])

mechanisms is more circumstantial. But at the least, there is now considerable evidence from a large number of quite disparate animals to suggest that neurosecretory mechanisms actually do have the general purpose of co-ordinating environmental change with developmental and physiological events within the animal. The environmental cues to which the neurosecretory system responds usually precede the situations which necessitate particular developmental and physiological changes. The animal is thus preadapted to take advantage of favourable or to avoid unfavourable conditions.

Of considerable interest in this connection is the part played by pheromones. These biologically very active molecules can be used to signal particular physiological states so that behavioural interactions will occur at appropriate times. The sex attractants of insects and other animals demonstrate the working of such a system. It is not surprising that both reproductive development and pheromone production are often coordinated by the endocrine system (p. 232).

But pheromones can also affect development itself. For example, the maturation pheromones of locusts control group reproduction, working either directly or indirectly through the endocrine systems of the recipients. The phenomenon is not confined to insects: in mice, for example, the male produces an olfactory pheromone which synchronizes oestrus cycles in groups of females. Moreover, the odour of a strange male can inhibit blastocyst implantation in recently mated females.[216] Here again, the pheromone is perceived by the recipient's olfactory receptors and ultimately acts through their endocrine systems. These pheromones may be considered as a kind of fine control over the effect produced by other internal and external forces. They assume overriding importance in the control of caste development in social insects.

Pheromones form one kind of chemical communication system in animals, hormones another. The effects of pheromones upon development illustrate how these two systems can interact. The metabolic machinery within a cell is a third kind of chemical communication system (p. 90), and there must be interaction between this and hormonal chemicals when the endocrine system controls metabolism and development. One way in which hormones can affect the intracellular system has already been described (p. 93). But hormones could act in many other ways: by altering membrane permeabilities, around and within the cell; by acting as co-enzymes; by activating pre-existing enzymes; and so on. It is likely that different hormones can affect the cell in various ways, and that they can have diverse effects upon different cells.

During embryogenesis, diffusable chemicals (organizer substances) are of the greatest importance in directing the development of particular cells and tissues along specific paths. In some animals, this process of determination may begin even in the unfertilized egg: in others, its commencement is delayed, even up to gastrulation. The fertilized egg possesses all the genes of the eventual adult individual. The progress of differentiation into cell layers, tissues and organs must involve the suppression of the activities of certain genes, and allow the full expression of others (p. 94). In those larval Diptera in which giant chromosomes can be seen in a number of tissues, the same arrangement of bands is everywhere recognizable. But chromosomal puffs, representing active genes, occur on different bands in different tissues (p. 93).

Differentiation continues in the post-embryonic stages in most animals, but is especially obvious in those with a metamorphosis from larval to adult form. It can be assumed that the sequence of gene suppressions and activations, initiated in the embryo, is continued into the post-embryonic period. But now hormones take over the job of the embryonic organizer substances: consequently, at whatever level hormones act upon the cell it is to be inferred that ultimately the genome itself is affected.

Factors other than hormones can influence the developmental capacity of the cell. For example, an insect epidermal cell has the capacity to form a gland cell, a sensory cell, a bristle cell or to remain as an ordinary epithelial cell.[280] The developmental course that the cell takes is determined, among other things, by the proximity of other such structures. These perhaps drain some essential material from the surrounding cells, or produce a substance which inhibits the development of like within a certain area.

Hormones are thus only one of a number of chemical communication systems which act and interact during the development of an adult, reproductive individual from the fertilized egg. As animals become structurally more complicated, the endocrine system assumes greater importance as an integrating and coordinating mechanism. But it is not impossible that many of the hormones present in the more complicated animals have their effects duplicated in the simpler ones by chemicals produced by, and acting within, individual cells and tissues. This is to say that increasing structural complexity, reflected in the greater specialization of individual cells for specific functions, necessitates the removal of some of the erstwhile more general properties of the cell to tissues designed for the purpose. Differentiation and development in a coelenterate cell proceeds as regularly and as smoothly as in a mammalian cell.

Once the embryonic period has been passed, differentiation and development must be related to environmental events for greatest efficiency. Consequently, at least one system for the translation of environmental stimuli into chemical messages must operate in all animals, and this system should exercise eventual control over the developmental machinery of the cell. It is suggested that the possession of neurosecretory mechanisms in animals as widely different in their structure and habits as coelenterates and man must reflect a common need to relate certain aspects of their development to particular environmental conditions.

References

1. ADKISSON, P. L. (1964). Action of the photoperiod in controlling insect diapause. *Am. Nat.* **98**, 357–374.
2. ANDREWARTHA, H. G. (1952). Diapause in relation to the ecology of insects. *Biol. Rev.* **27**, 50–107.
3. ANTHEUNISSE, L. J. (1963). Neurosecretory phenomena in the zebra mussel *Dreissena polymorpha* Pallas. *Archs Néerl. Zool.* **16**, 237–314.
4. AOTO, T. and NISHIDA, H. (1956). Effects of removal of the eyestalk on the growth and maturation of the oocytes in a hermaphrodite prawn, *Pandalus kessleri. J. Fac. Sci. Hokkaido Univ.* (Ser. 6. Zool.) **12**, 412–424.
5. BARRINGTON, E. W. J. (1964). Hormones and the control of colour. *The Hormones*, iv. Eds. PINCUS, G., THIMANN, K. V. and ASTWOOD, E. B. Academic Press, London and New York.
6. BARTH, R. H. (1961). Hormonal control of sex attractant in the Cuban cockroach. *Science, N.Y.* **133**, 1598–1599.
7. BARTH, R. H. (1965). Insect mating behaviour: endocrine control of a chemical communication system. *Science, N.Y.* **149**, 882–883.
8. BARTH, R. H. (1968). The comparative physiology of reproductive processes in cockroaches. *Advances in Reproductive Physiology*, vol. 3. Ed. MCLAREN, A. Logos Press, London.
9. BECKER, H. J. (1962). Die Puffs der Speicheldrüsenchromosomen von *Drosophila melanogaster*. II. Die Auslösung der Puffbildung, ihre Spezifität und ihre Beziehung zur funktion der Ringdrüse. *Chromosoma* (Berl.), **13**, 341–384.
10. BEERMANN, W. (1952). Chromomerenkonstanz und spezifische Modifikationen der Chromosomenstruktur in der Entwicklung und Organdifferenzierung von *Chironomus tentans. Chromosoma* (Berl.), **5**, 139–198.
11. BEERMANN, W. (1961). Ein Balbiani-ring als Locus einer Speicheldrüsenmutation. *Chromosoma* (Berl.), **12**, 1–25.

12. BELAMARICH, F. A. and TERWILLIGER, R. C. (1966). Isolation and identification of cardio-excitor hormone from the pericardial organs of *Cancer borealis*. *Am. Zool.* **6**, 101–106.

13. BERN, H. A. (1966). On the production of hormones by neurones and the role of neurosecretion in neuroendocrine mechanisms. *Symp. Soc. exp. Biol.* **20**, 325–344.

14. BERN, H. A., NISHIOKA, R. A. and HAGADORN, I. R. (1961). Association of elementary neurosecretory granules with the Golgi complex. *J. Ultrastruct. Res.* **5**, 311–320.

15. BERN, H. A. and TAKASUGI, N. (1962). The caudal neurosecretory system of fishes. *Gen. comp. Endocr.* **2**, 96–110.

16. BERN, H. A., YAGI, K. and NISHIOKA, R. S. (1965). Structure and function of the caudal neurosecretory system of fishes. *Arch. Anat. microsc.* **54**, 217–238.

17. BERRIDGE, M. J. (1966). The physiology of excretion in the cotton stainer, *Dysdercus fasciatus* Signoret. IV. Hormonal control of excretion. *J. exp. Biol.* **44**, 553–566.

17a. DE BESSÉ, N. (1967). Neurosécrétion dans la chaîne nerveuse ventrale de deux Blattes, *Leucophaea maderae* (F) et *Periplaneta americana* (L). *Bull. Soc. zool. Fr.* **92**, 73–86.

18. BLISS, D. E. (1951). Metabolic effects of sinus gland or eyestalk removal in the land crab, *Gecarcinus lateralis*. *Anat. Rec.* **111**, 502.

19. BLISS, D. E. (1953). Endocrine control of metabolism in the land crab *Gecarcinus lateralis* (Fréminville). I. Differences in respiratory metabolism of sinus glandless and eyestalkless crabs. *Biol. Bull.*, *Woods Hole*, **104**, 275–296.

20. BLISS, D. E., WANG, S. M. E. and MARTINEZ, E. A. (1966). Water balance in the land crab, *Gecarcinus lateralis*, during the intermoult cycle. *Am. Zool.* **6**, 197–212.

21. BOBIN, G. and DURCHON, M. (1953). Étude histologique du cerveau de *Perinereis cultrifera*. Mise en evidence d'une complèxe cérébro-vasculaire. *Archs Anat. microsc. Morph. exp.* **41**, 25–40.

22. BROWN, F. A., FINGERMAN, M. and HINES, M. N. (1952). Alterations in the capacity for light and dark adaptations of the distal retinal pigment of *Palaemonetes*. *Physiol. Zoöl.* **25**, 230–239.

23. BROWN, F. A., FINGERMAN, M., SANDEEN, M. I. and WEBB, H. M. (1953). Persistent diurnal and tidal rhythms of colour change in the fiddler crab, *Uca pugnax*. *J. exp. Zool.* **123**, 29.

24. BROWN, F. A., HINES, M. N. and FINGERMAN, M. (1952). Hormonal regulation of the distal retinal pigment of *Palaemonetes*. *Biol. Bull.*, *Woods Hole*, **102**, 212–225.

25. BROWN, F. A. and SANDEEN, M. I. (1948). Responses of the chromatophores of the fiddler crab, *Uca*, to light and temperature. *Physiol. Zoöl.* **21**, 361.

26. BROWN, F. A., WEBB, H. M. and SANDEEN, M. I. (1952). The action of two hormones regulating the red chromatophores of *Palaemonetes*. *J. exp. Zool.* **120**, 391.

27. BROWN, F. A., WEBB, H. M. and SANDEEN, M. (1953). Differential production of two retinal pigment hormones in *Palaemonetes* by light flashes. *J. cell. comp. Physiol.* **41**, 123–144.

28. BROWN, F. A. and WULFF, V. J. (1941). Chromatophore types in *Crangon* and their endocrine control. *J. cell. comp. Physiol.* **18**, 339–353.

29. BURNETT, A. L. and DIEHL, N. A. (1964). The nervous system of *Hydra*. I. Types, distribution and origin of nerve elements. *J. exp. Zool.* **157**, 217–226.

30. BURNETT, A. L., DIEHL, N. A. and DIEHL, F. (1964). The nervous system of *Hydra*. II. Control of growth and regeneration by neurosecretory cells. *J. exp. Zool.* **157**, 227–236.

31. BURNETT, A. L. and DIEHL, N. A. (1964). The nervous system of *Hydra*. III. The initiation of sexuality with special reference to the nervous system. *J. exp. Zool.* **157**, 237–250.

32. BUTENANDT, A. (1963). Bombycol, the sex-attractive substance of the silkworm, *Bombyx mori*. *J. Endocr.* **27**, ix–xvi.

33. BUTENANDT, A. and KARLSON, P. (1954). Über die Isolierung eines Metamorphose-Hormons der Insekten in kristellisierter Form. *Z. Naturf.* **96**, 389–391.

34. BUTLER, C. G. (1967). Insect pheromones. *Biol. Rev.* **42**, 42–87.

35. BUTLER, C. G. and CALLOW, R. K. (1968). Pheromones of the honey bee (*Apis mellifera* L): the 'inhibitory scent' of the queen. *Proc. R. ent. Soc. Lond.* **43**, 62–65.

36. CARLISLE, D. B. (1953). Moulting hormones in *Leander* (Crustacea Decapoda). *J. mar. biol. Ass. U.K.* **32**, 289–296.

37. CARLISLE, D. B. (1953). Studies on *Lysmata seticaudata* Risso (Crustacea Decapoda). V. The ovarian inhibiting hormone and the hormonal inhibition of sex reversal. *Pubbl. Staz. zool. Napoli*, **24**, 355–372.

38. CARLISLE, D. B. (1956). On the hormonal control of water balance in *Carcinus*. *Pubbl. Staz. zool. Napoli*, **27**, 227–231.

39. CARLISLE, D. B. (1957). On the hormonal inhibition of moulting in decapod crustacea. II. The terminal anecdysis in crabs. *J. mar. biol. Ass. U.K.* **36**, 291–307.

40. CARLISLE, D. B. and DOHRN, P. F. R. (1952). Sulla presenza di un ormone d'accresciento in un crostaceo decapoda, la *Lysmata seticaudata* Risso. *Ric. sci. Torino*, **23**, 95–100.

41. CARLISLE, D. B. and KNOWLES, F. (1959). *Endocrine control in crustaceans.* Cambridge University Press.

42. CASANOVA, G. (1955). Influence du prostomium sur la régénération caudale chez *Platynereis massiliensis* (Moquin-Tandon). *C.r. Acad. Sci. Paris*, **240**, 1814–1816.

43. CAZAL, P. (1948). Les glandes endocrines rétrocérébrales des insectes (étude morphologique). *Bull. biol. Fr. Belg.* (Suppl.), **32**, 1–227.

44. CHAET, A. B. (1966). The gamete-shedding substances of starfishes: a physiological biochemical study. *Am. Zool.* **6**, 263–271.

44a. CHALAYE, D. (1967). Neurosécrétion au niveau de la chaîne nerveuse ventrale de *Locusta migratoria migratorioides* R et F. (Orthoptère Acridien). *Bull. Soc. Zool. Fr.* **92**, 87–108.

45. CHARNIAUX-COTTON, H. (1952). Castration chirurgicale chez un Crustacé Amphipode (*Orchestia gammarella*) et déterminisme des caractères sexuel secondaires. Premier résultats. *C.r. Acad. Sci. Paris*, **234**, 2570–2572.

46. CHARNIAUX-COTTON, H. (1954). Découverte chez un crustacé amphipode (*Orchestia gammarella*) d'une gland endocrine responsible de la differentiation de caractères sexuel primaires et sécondaires mâle. *C.r. Acad. Sci. Paris*, **239**, 780–782.

47. CHARNIAUX-COTTON, H. (1956). Existance d'un organ comparable à la 'glande androgène' chez un Pagure et un crabe. *C.r. Acad. Sci. Paris*, **243**, 1487–1489.
48. CHARNIAUX-COTTON, H. (1957). Croissance, régénération et déterminisme endocrinien des caractères sexuels d'*Orchestia gammarella* (Pallas) Crustacé Amphipode. *Ann. sci. nat. Zool. et biol. animale*, **19**, 411–559.
49. CHARNIAUX-COTTON, H. (1958). La glande androgène de quelques Crustacés Décapodes et particulièrement de *Lysmata seticaudata* espèce à hermaphrodisme protérandrique fonctionnel. *C.r. Acad. Sci. Paris*, **246**, 2817–2819.
50. CHARNIAUX-COTTON, H. (1958). Contrôle de la différenciation du sexe et de la reproduction chez les Crustacés supérieurs. *Bull. Soc. zool. Fr.* **83**, 314–336.
51. CHARNIAUX-COTTON, H. (1960). Sex determination. *Physiology of Crustacea* I. Ed. WATERMAN, T. H. Academic Press, London and New York.
52. CLARK, R. B. (1956). On the origin of neurosecretory cells. *Ann. Sci. Nat. Zool. et biol. Animale*. **18**, 199–207.
53. CLARK, R. B. (1958). The micromorphology of the supra-oesophageal ganglion of *Nephtys*. *Zool. Jb.* (*Physiol.*), **68**, 261–299.
54. CLARK, R. B. (1959). The neurosecretory system of the supra-oesophageal ganglion of *Nephtys*. *Zool. Jb.* (*Physiol*), **68**, 395–424.
55. CLARK, R. B. (1961). The origin and formation of the heteronereis. *Biol. Rev.* **36**, 199–236.
56. CLARK, R. B. (1965). Endocrinology and reproductive biology of polychaetes. *Oceanography and Marine Biology, An Annual Review* **3**, Ed. BARNES, H. Allen and Unwin, London.
57. CLARK, R. B. and BONNEY, D. G. (1960). Influence of the supra-oesophageal ganglion on posterior regeneration in *Nereis diversicolor*. *J. Embryol. exp. Morph.* **8**, 112–118.
58. CLARK, R. B. and EVANS, S. M. (1961). The effects of delayed brain extirpation and replacement on caudal regeneration in *Nereis diversicolor*. *J. Embryol. exp. Morph.* **9**, 97–105.
59. CLEVER, U. (1961). Genaktivitäten in den Reisenchromosomen von *Chironomus tentans* und ihre Beziehungen zur Entwicklung. I. Genaktivierungen durch Ecdyson. *Chromosoma* (Berl.), **12**, 607–675.
60. CLEVER, U. (1962). Genaktivitäten in den Riesenchromosomen von *Chironomus tentans* und ihre Beziehungen zur Entwicklung. II. Das verhalten der Puffs während des letzten Larvenstadiums und der Puppenhäutung. *Chromosoma* (Berl.), **13**, 385–486.
61. CLEVER, U. (1964). Actinomycin and puromycin: effects on sequential gene activation by ecdysone. *Science* N.Y., **146**, 794–795.
62. COLES, G. C. (1965). Haemolymph proteins and yolk formation in *Rhodnius prolixus*. Stal. *J. exp. Biol.* **43**, 425–431.
63. COLES, G. C. (1965). Studies on the hormonal control of metabolism in *Rhodnius prolixus* Stal. I. The adult female. *J. Insect Physiol.* **11**, 1325–1330.
64. COOKE, I. M. (1966). The site of action of pericardial organ extract and 5-hydroxytrypamine in the Decapod Crustacean heart. *Am. Zool.* **6**, 107–121.

65. DAHM, K. H., ROLLER, H. and TROST, B. M. (1968). The juvenile hormone. IV. Stereochemistry of juvenile hormone and biological activity of some of its isomers and related compounds. *Life Sci.* **7**, 129–138.

66. DANIEL, P. M. (1966). Blood supply of hypothalamus and pituitary gland. *Br. med. Bull.* **22**, 202–208.

67. DANILEVSKII, A. S. (1965). *Photoperiodism and seasonal development of insects.* Oliver and Boyd, Edinburgh and London.

68. DAVEY, K. G. (1958). The migration of spermatozoa in the female of *Rhodnius prolixus* Stal. *J. exp. Biol.* **35**, 694–701.

69. DAVEY, K. G. (1959). Spermatophore production in *Rhodnius prolixus. Q. Jl. microsc. Sci.* **100**, 221–230.

70. DAVEY, K. G. (1961). The mode of action of the heart accelerating factor from the corpus cardiacum of insects. *Gen. comp. Endocr.* **1**, 24–29.

71. DAVEY, K. G. (1961). The release by feeding of a pharmacologically active factor from the corpus cardiacum of *Periplaneta americana. J. Insect Physiol.* **8**, 205–208.

72. DAVEY, K. G. (1962). The nervous pathway involved in the release by feeding of a pharmacologically active factor from the corpus cardiacum of *Periplaneta. J. Insect Physiol.* **8**, 579–583.

73. DAVEY, K. G. (1965). *Reproduction in the insects.* Oliver and Boyd, Edinburgh and London.

74. DAVEY, K. G. (1966). Neurosecretion and moulting in some parasitic nematodes. *Am. Zool.* **6**, 243–250.

75. DAVEY, K. G. (1967). Some consequences of copulation in *Rhodnius prolixus. J. Insect Physiol.* **13**, 1629–1636.

76. DELPHIN, G. (1965). The histology and possible functions of neuro-secretory cells in the ventral ganglia of *Schistocerca gregaria* Forskål (Orthoptera: Acrididae). *Trans. R. ent. Soc. Lond.* **117**, 167–214.

77. DEMEUSY, N. (1958). Recherches sur la mue de puberté du decapode brachyoure *Carcinus maenas* Linné. *Archs Zool. exp. gén.* **95**, 253–491.

77a. DUBOIS, F. S. and LENDER, T. (1956). Corrélations humorales dans la régenération des planaires paludicoles. *Ann. Sci. Nat. Zool. et biol. Animale.* **18**, 223–230

78. DUPONT-RAABE, M. (1966). Étude des phénomènes de neurosécrétion au niveau de la chaîne nerveuse ventrale des phasmides. *Bull. Soc. zool. Fr.* **90**, 631–654.

79. DURCHON, M. (1956). Influence du cerveau sur les processus de régénération caudale chez Néréidiens (Annélides Polychètes). *Archs Zool. exp. gén.* **94** (N and R), 1–9.

80. DURCHON, H. (1960). L'endocrinologie chez les annélides polychètes. *Bull. Soc. zool. Fr.* **85**, 275–301.

81. DURCHON, M., MONTREUIL, J. and BOILLY-MARER, Y. (1963). Résultat préliminaires sur la nature chimique de l'hormone inhibitrice du cerveau des Néréidiens (Annélides Polychètes). *C.r. Acad. Sci. Paris*, **257**, 1807–1808.

82. ECHALIER, G. (1954). Recherches expérimèntales sur le rôle de 'l'organe Y' dans la mue de *Carcinus maenas* (L). Crustacé decapode. *C.r. Acad. Sci. Paris*, **238**, 523–525.

83. ECHALIER, G. (1955). Rôle de l'organe Y dans la déterminisme de la mue de *Carcinides* (*Carcinus*) *maenas* L. (Crustacé decapoda). Expérience d'implantation. *C.r. Acad. Sci., Paris*, **240**, 1581–1583.

84. EDSTROM, J. E. and BEERMANN, W. (1962). The base composition of nucleic acids in chromosomes, puffs, nucleoli and cytoplasm of *Chironomus* salivary gland cells. *J. cell. Biol.* **14**, 371–379.

85. ENGELMANN, F. (1960). Mechanisms controlling reproduction in two viviparous roaches (Blattaria). *Ann. N.Y. Acad. Sci.* **89**, 516–536.

86. ENGELMANN, F. (1965). The mode of regulation of the corpus allatum in adult insects. *Archs Anat. microsc. Morph. exp.* **54**, 387–404.

87. FINGERMAN, M. (1959). Comparison of the chromatophorotropins of two crayfish with special reference to electrophoretic behaviour. *Tulane Stud. Zool.* **7**, 21.

88. FINGERMAN, M. (1963). *The control of chromatophores.* Pergamon Press, Oxford.

89. FINGERMAN, M. (1966). Neurosecretory control of pigmentary effectors in crustaceans. *Am. Zool.* **6**, 169–179.

90. FINGERMAN, M. and AOTA, T. (1958). Electrophoretic analysis of chromatophorotropins in the dwarf crayfish *Cambarellus shufeldti*. *J. exp. Zool.* **138**, 25.

91. FINGERMAN, M., DOMINIEZAK, T., MIYAWAKI, M. and OGURO, C. (1967). Neuroendocrine control of the hepatopancreas in the crayfish *Procambarus clarki*. *Physiol. Zoöl.* **40**, 23–30.

92. FINGERMAN, M., SANDEEN, M. I. and LOWE, M. E. (1958). Experimental analysis of the red chromatophore system of the prawn *Palaemonetes vulgaris*. *Physiol. Zoöl.* **32**, 128–149.

93. FINGERMAN, M. and TINKLE, D. W. (1956). Responses of the white chromatophores of two species of prawns (*Palaemonetes*) to light and temperature. *Biol. Bull., Woods Hole*, **110**, 144.

94. FRAENKEL, G. (1935). A hormone causing pupation in the blowfly, *Calliphora erythrocephala*. *Proc. R. Soc.* (B), **118**, 1–12.

95. FRAENKEL, G. and HSIAO, C. (1965). Bursicon, a hormone which mediates tanning of the cuticle in the adult fly and other insects. *J. Insect. Physiol.* **11**, 513–526.

96. FUKAYA, M. (1962). The inhibitory action of farnesol on the development of the rice stem borer in post-diapause. *Jap. J. appl. Ent. Zool.* **6**, 298.

97. FUKUDA, S. (1963). Déterminisme hormonal de la diapause chez le ver á soie. *Bull. Soc. zool. Fr.* **88**, 151–179.

98. GABE, M. (1955). Données histologiques sur la neurosécrétion chez les Arachnides. *Archs Anat. microsc.* **44**, 351–383.

99. GABE, M. (1956). Histologie comparé de la glande de mue (organe Y) des Crustacés malacostracés. *Ann. Sci. nat.* (*Zool.*), **18**, 145–152.

100. GABE, M. (1966). *Neurosecretion.* Pergamon Press, Oxford.

101. GALBRAITH, M. N., HORN, D. H. S., MIDDLETON, E. J. and HACKNEY, R. J. (1968). Structure of deoxycrustecdysone, a second crustacean moulting hormone. *Chem. Comm.* **2**, 83–85.

102. GALLISSIAN, A. (1963). Action des ganglions cérèbroîdes sur la diapause et la régénération d'*Eophila dollfusa* Tetry. (Lumbricide). *C.r. Acad. Sci. Paris*, **256**, 1158.

103. GANGULY, D. N. and RAY, A. K. (1961). Caudal neurosecretory system in the vertebrates. *Sci. Cult.* **27**, 585–586.

104. GELDIAY, S. (1967). Hormonal control of adult reproductive diapause in the Egyptian grasshopper, *Anacridium aegyptium* L. *J. Endocr.* **37**, 63–71.

105. GERSCH, M. (1961). Insect metamorphosis and the activation hormone. *Am. Zool.* **1**, 53–57.

106. GERSCH, M. (1962). The activation hormone of the metamorphosis of insects. *Gen. comp. Endocr.* Suppl. **1**, 322–329.

107. GILBERT, L. I. and GOODFELLOW, R. D. (1965). Endocrinological significance of sterols and isoprenoids in the metamorphosis of the American Silkmoth *Hyalophora cecropia*. *Zool. Jb.* (*Physiol.*), **8**, 718–726.

108. GILBERT, L. I. and SCHNEIDERMAN, H. A. (1959). Prothoracic gland stimulation by juvenile hormone extracts of insects. *Nature, Lond.* **184**, 171–173.

109. GILBERT, L. I. and SCHNEIDERMAN, H. A. (1960). The development of a bioassay for the juvenile hormone of insects. *Trans. Am. microsc. Soc.* **74**, 38–67.

110. GILBERT, L. I. and SCHNEIDERMAN, H. A. (1961). The content of juvenile hormone and lipid in Lepidoptera: sexual differences and developmental changes. *Gen. comp. Endocr.* **1**, 453–472.

111. GILGAN, M. W. and IDLER, D. R. (1967). The conversion of androstenedione to testosterone by some lobster (*Homarus americanus* Milne Edwards) tissues. *Gen. comp. Endocr.* **9**, 319–324.

112. GIRARDIE, A. (1964). Action de la pars intercérèbralis sur le dévelopment de *Locusta migratoria* L. *J. Insect Physiol.* **10**, 599–609.

113. GIRARDIE, A. (1967). Controle neuro-hormonal de la métamorphose et de la pigmentation chez *Locusta migratoria cinerascens* (Orthoptère). *Bull. biol. Fr. Belg.* **101**, 79–114.

114. GOLDING, D. W. (1967). Regeneration and growth control in *Nereis*. I. Growth and regeneration. *J. Embryol. exp. Morph.* **18**, 67–77.

115. GOLDING, D. W. (1967). Regeneration and growth control in *Nereis*. II. An axial gradient in growth potentiality. *J. Embryol. exp. Morph.* **18**, 79–90.

116. GOMEZ, R. (1965). Acceleration of development of the gonads by implantation of brain in the crab *Paratelphusa hydrodromus*. *Naturwissenschaften*, **52**, 216.

117. GOODFELLOW, R. D. and GILBERT, L. I. (1963). Sterols and terpenes in the Cecropia silkmoth. *Am. Zool.* **3**, 137.

118. GORBMAN, A. and BERN, H. A. (1962). *A textbook of comparative endocrinology*. John Wiley, New York.

119. HAGADORN, I. R. (1958). Neurosecretion and the brain of the rhynchobdellid leech *Theromyzon rude*. (Baird, 1869). *J. Morph.* **102**, 55–90.

120. HAGADORN, I. R. (1966). Neurosecretion in the Hirudinea and its possible role in reproduction. *Am. Zool.* **6**, 251–262.

121. HAMPSHIRE, F. and HORN, D. H. S. (1966). *Chem. Comm.* **2**, 37.

122. HANSTRÖM, B. (1954). On the transformation of ordinary nerve cells into neurosecretory cells. *K. fysiogr. Sällsk. Lund. Förh.* **24**, 1–8.

123. HARRIS, G. W., REED, M. and FAWCETT, C. P. (1966). Hypothalamic releasing factors and the control of anterior pituitary function. *Br. med. Bull.* **22**, 266–272.

124. HAUENSCHILD, C. (1960). Lunar periodicity. *Cold Spring Harb. Symp. quant. Biol.* **25**, 491–497.

125. HAUENSCHILD, C. (1964). Postembryonale Entwicklungssteuerung durch ein Gehirn-Hormon bei *Platynereis dumerilii*. *Zool. Anz.* Suppl. **27**, 111–120.

254   REFERENCES

126. HAUENSCHILD, C. and FISCHER, A. (1962). Neurosecretory control of development in *Platynereis dumerilii*. *Mem. Soc. Endocr.* **12**, 297–312.

127. HERLANT-MEEWIS, H. (1957). Réproduction et neurosécrétion chez *Eisenia foetida* (Sav). *Annls Soc. r. zool. Belg.* **87**, 151–183.

128. HERLANT-MEEWIS, H. (1959). Phénomènes neurosécrétoire et sexualité chez *Eisenia foetida*. *C.r. Acad. Sci. Paris*, **248**, 1405–1407.

129. HERLANT-MEEWIS, H. (1964). Regeneration in annelids. *Advances in Morphogenesis*, vol. 4. Eds. ABERCROMBIE, M. and BRACHET, J. Academic Press, London and New York.

130. HEYMONS, R. (1895). Die Embryonalentwicklung von Dermapteren und Orthopteren, unter besonderer Berücksichtigung der Keimblätterbildung. Jena.

131. HIGHNAM, K. C. (1961). The histology of the neurosecretory system of the adult female desert locust, *Schistocerca gregaria*. *Q. Jl. microsc. Sci.* **102**, 27–38.

132. HIGHNAM, K. C. (1961). Induced changes in the amounts of material in the neurosecretory system of the desert locust. *Nature, Lond.* **191**, 199–200.

133. HIGHNAM, K. C. (1962). Neurosecretory control of ovarian development in *Schistocerca gregaria*. *Q. Jl. microsc. Sci.* **103**, 57–72.

134. HIGHNAM, K. C. (1964). Endocrine relationships in insect reproduction. In *Insect Reproduction*, pp. 26–42. Symp. No. 2. Royal Ent. Soc. Lond.

135. HIGHNAM, K. C. (1964). Hormones and behaviour in insects. *Viewpoints in Biology*, **3**, 219–260.

136. HIGHNAM, K. C. (1965). Some aspects of neurosecretion in arthropods. *Zool. Jb. (Physiol.)*, **71**, 558–582.

137. HIGHNAM, K. C. (1969). Estimates of neurosecretory activity during maturation in locusts. *Proc. Int. Symp. Ins. Endocr. Brno.* (In press.)

138. HIGHNAM, K. C. and HASKELL, P. T. (1964). The endocrine systems of isolated and crowded *Locusta* and *Schistocerca* in relation to oocyte growth, and the effects of flying upon maturation. *J. Insect Physiol.* **10**, 849–864.

139. HIGHNAM, K. C., HILL, L. and GINGELL, D. (1965). Neurosecretion and water balance in the male desert locust (*Schistocerca gregaria*). *J. Zool.* **147**, 201–215.

140. HIGHNAM, K. C., HILL, L. and MORDUE, W. (1966). The endocrine system and oocyte growth in *Schistocerca* in relation to starvation and frontal ganglionectomy. *J. Insect Physiol.* **12**, 977–994.

141. HIGHNAM, K. C. and LUSIS, O. (1962). The effect of mature males on the neurosecretory control of ovarian development in the desert locust. *Q. Jl. microsc. Sci.* **103**, 73–83.

142. HIGHNAM, K. C., LUSIS, O. and HILL, L. (1963). The role of the corpora allata during oocyte growth in the desert locust, *Schistocerca gregaria* Forsk. *J. Insect Physiol.* **9**, 587–596.

143. HIGHNAM, K. C., LUSIS, O. and HILL, L. (1963). Factors affecting oocyte resorption in the desert locust, *Schistocerca gregaria*. *J. Insect Physiol.* **9**, 827–837.

144. HILL, L. (1962). Neurosecretory control of haemolymph protein concentration during ovarian development in the desert locust. *J. Insect Physiol.* **8**, 609–619.

145. HILL, L. (1965). The incorporation of C^{14}-glycine into the proteins of the fatbody of the desert locust during ovarian development. *J. Insect Physiol.* **11**, 1605–1615.

146. HILL, L. and GOLDSWORTHY, G. J. (1968). Growth, feeding activity and the utilization of reserves in larvae of *Locusta*. *J. Insect. Physiol.* **14**, 1085–1098.

147. HILL, L., LUNTZ, A. J. and STEELE, P. A. (1968). The relationships between somatic growth, ovarian growth, and feeding activity in the adult desert locust. *J. Insect Physiol.* **14**, 1–20.

148. HILL, L., MORDUE, W. and HIGHNAM, K. C. (1966). The endocrine system, frontal ganglion, and feeding during maturation in the female desert locust. *J. Insect Physiol.* **12**, 1197–1208.

149. HOFFMEISTER, H. (1966). Ecdysteron, ein neues Häutungshormon der Insekten. *Angew. Chem.* **78**, 269–270.

150. HOFFMEISTER, H., NAKANISHI, K., KOREEDA, M. and HSU, H. Y. (1968). The moulting hormone activity of ponasterones in the *Calliphora* test. *J. Insect Physiol.* **14**, 53–54.

151. HOWIE, D. I. D. (1963). Experimental evidence for the humoral stimulation of ripening of the gametes and spawning in the polychaet *Arenicola marina* (L). *Gen. comp. Endocr.* **3**, 660–668.

152. ICHIKAWA, M. (1962). Brain and metamorphosis of Lepidoptera. *Gen. comp. Endocr.* Suppl. **1**, 331–336.

153. ICHIKAWA, M. and ISHIZAKI, H. (1961). Brain hormone of the silkworm, *Bombyx mori*. *Nature, Lond.* **191**, 933–934.

154. ITO, H. (1918). On the glandular nature of the Corpora allata of the Lepidoptera. *Bull. imp. seric. Coll., Tokyo*, **1**, 64–103.

155. JACOBSON, M., BEROZA, M. and JONES, W. A. (1960). Isolation, identification and synthesis of the sex attractant of the gypsy moth. *Science N.Y.* **139**, 48–49.

156. JOHNSON, B. (1963). A histological study of neurosecretion in aphids. *J. Insect Physiol.* **9**, 727–739.

157. JOLY, P. (1945). La fonction ovarienne et son contrôle humorale chez les dytiscides. *Archs. Zool. exp. gén.* **84**, 47–164.

158. JOOSSE, J. (1964). Dorsal bodies and dorsal neurosecretory cells of the cerebral ganglia of *Lymnaea stagnalis*. *Archs. Néerl. Zool.* **16**, 1–103.

159. KAPLANIS, J. N., THOMPSON, M. J., YAMAMOTO, R. T., ROBBINS, W. E. and LOULOUDES, S. J. (1966). Ecdysones from the pupa of the tobacco hornworm, *Manduca sexta* (Johannson). *Steroids*, **8**, 605–623.

160. KARLSON, P. (1963). Chemie und Biochemie der Insektenhormone. *Angew. Chem.* **75**, 257–265.

161. KARLSON, P. and HOFFMEISTER, H. (1963). Zur Biogenese des Ecdysons. I. Umwandlung von Cholesterin in Ecdyson. *Hoppe-Seylers. Z. physiol. Chem.* **331**, 298–300.

162. KARLSON, P., HOFFMEISTER, H., HOPPE, W. and HUBER, R. (1963). Zur Chemie des Ecdysons. *Justus Liebigs Annl. or. Chem.* **662**, 1–20.

163. KARLSON, P. and LUSCHER, M. (1959). 'Pheromones': a new term for a class of biologically active substances. *Nature, Lond.* **183**, 55–56.

164. KARLSON, P. and SCHMIALEK, H. (1959). Nachreis der Exkretion von Juvenilhormon. *Z. Naturf.* **146**, 821.

165. KARLSON, P. and SCHWEIGER, A. (1961). Zum Tyrosinstoffwechsel der Insekten. IV. Mitteilung Das Phenoloxydase-System von *Calliphora* und seine Beeinflussung durch das Hormon Ecdyson. *Hoppe-Seylers Z. physiol. Chem.* **323**, 199–210.

166. KARLSON, P. and SEKERIS, C. (1962). Zum Tyrosinstoffwechsel der Insekten IX. Kontrolle des Tyrosinstoffwechsels durch Ecdyson. *Biochim. biophys. Acta.* **63**, 489–495.

167. KARLSON, P. and SEKERIS, C. (1966). Ecdysone, an insect steroid hormone, and its mode of action. *Rec. Progr. Horm. Res.* **22**, 473–502.

168. KENNEDY, J. S. (1965). Coordination of successive activities in an aphid. Reciprocal effects of settling on flight. *J. exp. Biol.* **43**, 489–509.

169. KENNEDY, J. S. (1967). Behaviour as physiology. *Insects and physiology.* Eds. BEAMENT, J. W. L. and TREHERNE, J. E. Oliver and Boyd, Edinburgh and London. 1967.

170. KLEINHOLZ, L. H. (1936). Crustacean eyestalk hormones and retinal pigment migration. *Biol. Bull., Woods Hole,* **70**, 159–184.

171. KLEINHOLZ, L. H. (1961). Pigmentary effectors. *Physiology of Crustacea.* II. Ed. WATERMAN, T. H. Academic Press, London and New York.

172. KNOWLES, F. G. W., CARLISLE, D. B. and DUPONT-RAABE, M. (1956). Inactivation enzymatique d'une substance chromactive des insectes et des crustacés. *C.r. Acad. Sci. Paris,* **242**, 825.

173. KOBAYASHI, M. and KIRIMURA, J. (1958). The 'brain' hormone in the silkworm, *Bombyx mori. Nature,* Lond. **181**, 1217.

174. KOBAYASHI, M., KIRIMURA, J. and SAITO, M. (1962). The 'brain' hormone in an insect *Bombyx mori* L. (Lepidoptera). *Mushi,* **36**, 85–92.

175. KOBAYASHI, M., TAKEMOTO, T., OGAWA, S. and NISHIMOTO, N. (1967). The moulting hormone activity of ecdysterone and inokosterone isolated from *Achyranthis radix. J. Insect Physiol.* **13**, 1395–1399.

176. KOLLER, G. (1928). Versuch über die inkretorischen Vorgange bien Garneelfarbweichsel. *Z. vergl. Physiol.* **8**, 601.

177. KOPEC, S. (1922). Studies on the necessity of the brain for the inception of insect metamorphosis. *Biol. Bull., Woods Hole,* **42**, 323–342.

178. KRISHNAKUMARAN, A., OBERLANDER, H. and SCHNEIDERMAN, H. A. (1965). Rates of DNA and RNA synthesis in various tissues during a larval moult cycle of *Samia cynthia ricini* (Lepidoptera). *Nature, Lond.* **205**, 1131–1133.

179. KURUP, N. G. and SCHEER, B. T. (1966). Control of protein synthesis in an anomuran crustacean. *Comp. Biochem. Physiol.* **18**, 971.

180. LANE, N. J. (1962). Neurosecretory cells in the optic tentacles of certain pulmonates. *Q. Jl. microsc. Sci.* **103**, 211–223.

181. LANE, N. J. (1964). Elementary neurosecretory granules in neurones of the snail, *Helix aspera. Q. Jl. microsc. Sci.* **105**, 31–34.

182. LAUFER, H. (1965). Developmental studies of the Dipteran salivary gland. III. Relationships between chromosomal puffing and cellular function during development. *Developmental and metabolic control mechanisms and Neoplasia.* Williams & Wilkins Co., Baltimore.

183. LAUFER, H. and GOLDSMITH, M. (1965). Ultrastructural evidence for a protein transport system in *Chironomus* salivary glands and its implications for chromosomal puffing. *J. cell. Biol.* **27**, 57A.

184. LAUFER, H. and NAKASE, Y. (1965). Developmental studies of the Dipteran salivary gland. II. DNAase activity in *Chironomus thummi. J. cell. Biol.* **25**, 97–102.

185. LAUFER, H. and NAKASE, Y. (1965). Salivary gland secretion and its relation to chromosomal puffing in the Dipteran *Chironomus thummi*. *Proc. natn. Acad. Sci.* **53**, 511–516.

186. LECHENAULT, H. (1962). Sur l'existence de cellules neurosécrétrices dans les ganglions cérébroides des Lineidae (Hétéronémertes). *C.r. Acad. Sci. Paris*, **255**, 194–196.

187. LEES, A. D. (1955). *The physiology of diapause in arthropods*. Cambridge University Press.

188. LEGENDRE, R. (1959). Contribution à l'étude du systéme nerveux des Aranéides. *Ann. Sci. nat. Zool.* **1**, 339–474.

189. LENDER, TH. and KLEIN, N. (1961). Misé en evidence de cellules sécrétrices dans le cerveau de la Planaire *Polycelis nigra*. Variations de leur nombre au cours de la regénération postérieure. *C.r. Acad. Sci. Paris*, **253**, 331–333.

190. LEVER, J., DE VRIES, C. M. and JAGER, J. C. (1965). On the anatomy of the central nervous system and the location of neurosecretory cells in *Australorbis glabratus. Malacologia*, **2**, 219–230.

191. LEVER, J., JANSEN, J. and DE VLIEGER, T. A. (1961). Pleural ganglia and water balance in the freshwater pulmonate *Lymnaea stagnalis. Proc. K. ned. Akad. Wet.* Section C. **64**, 532–542.

192. LOHER, W. (1960). The chemical acceleration of the maturation process and its hormonal control in the male of the desert locust. *Proc. R. Soc.* (B), **153**, 380–397.

193. LOHER, W. and HUBER, F. (1966). Nervous and endocrine control of sexual behaviour in a grasshopper (*Gomphocerus rufus* L., Acridinae). *Symp. Soc. exp. Biol.* **20**, 381–400.

194. LUBET, P. (1955). Cycle neurosécrétroire de *Chlamys varia* L. et *Mytilus edulis* L. *C.r. Acad. Sci. Paris*, **241**, 119–121.

195. LUBET, P. (1956). Effects de l'ablation des centres nerveux sur l'émission des gametes chez *Mytilus edulis* L. et *Chlamys varia* L. *Ann. Sci. natur. Zool.* **18**, 175–183.

196. LÜSCHER, M. (1962). In *Insect Polymorphism*. Social control of polymorphism in termites. Ed. KENNEDY, J. S. Symp. No. **1**. Royal Ent. Soc. Lond. 57–67.

197. LÜSCHER, M. and KARLSON, P. (1958). Experimentelle Auslösung von Häutungen bei der Termite *Kalotermes flavicollis. J. Insect Physiol.* **1**, 341–345.

198. LUSIS, O. (1963). The histology and histochemistry of development and resorption in the terminal oocytes of the desert locust *Schistocerca gregaria. Q. Jl. microsc. Sci.* **104**, 57–68.

199. MADDRELL, S. H. P. (1963). Excretion in the blood-sucking bug, *Rhodnius prolixus* Stal. 1. The control of diuresis. *J. exp. Biol.* **40**, 247–256.

200. MADHAVAN, K. and SCHNEIDERMAN, H. A. (1968). Effects of ecdysone on epidermal cells in which DNA synthesis has been blocked. *J. Insect Physiol.* **14**, 777–781.

201. MANNING, A. (1967). *An introduction to animal behaviour*. Edward Arnold, London.

202. MASNER, P., SLAMA, K. and LANDA, V. (1968). Sexually spread insect sterility induced by the analogues of juvenile hormone. *Nature, Lond.* **219**, 395–396.

203. MAYNARD, D. M. (1960). Circulation and heart function. *Physiology of Crustacea*, I. Ed. WATERMAN, T. H. Academic Press, London and New York.

204. MCWHINNIE, M. A. and CHUA, A. (1964). Hormonal regulation of crustacean tissue metabolism. *Gen. comp. Endocr.* **4**, 624–633.

205. MCWHINNIE, M. A. and CHUA, A. (1964). Eyestalk influence on carbohydrate metabolism in the crayfish *Orconectes virilis*. *Am. Zool.* **4**, 298.

206. MECHELKE, F. (1953). Reversible Strukturmodificationen der Speicheldrüsenchromosomen von *Acricotopus lucidus*. *Chromosoma* (Berl.), **5**, 511.

207. MILLS, R. R. and NEILSEN, D. J. (1967). Hormonal control of tanning in the American cockroach. V. Some properties of the purified hormone. *J. Insect Physiol.* **13**, 273–280.

208. MINKS, A. (1967). Biochemical aspects of juvenile hormone action in the adult *Locusta migratoria*. *Arch. Néerl. Zool.* **17**, 175–257.

209. MORDUE, W. (1969). Hormonal control of Malpighian tube and rectal function in the desert locust, *Schistocerca gregaria*. *J. Insect Physiol.* **15**, 273–285.

210. MORDUE, W. and GOLDSWORTHY, G. J. (1969). The physiological effects of corpus cardiacum extracts in locusts. *Gen. comp. Endocr.* **12**, 360–369.

211. NAISSE, J. (1965). Contrôle endocrinien de la différenciation sexuelle chez les insectes. *Arch. d'anat. micr. et exper.* **54**, 417–428.

212. NAYAR, K. K. (1958). Studies on the neurosecretory system of *Iphita limbata* Stål. V. Probable endocrine basis of oviposition in the female insect. *Proc. Indian Acad. Sci.* **47**, 233–251.

213. NELSON, J. A. (1915). *The embryology of the honey bee*. Princeton University Press.

214. ODHIAMBO, T. R. (1966). The fine structure of the corpus allatum of the sexually mature male of the desert locust. *J. Insect Physiol.* **12**, 819–828.

215. PANOUSE, J. B. (1943). Influence de l'ablation du pedoncule oculaire sur la croissance de l'ovaire chez la crevette *Leander serratus*. *C.r. Acad. Sci. Paris*, **217**, 553–555.

216. PARKES, A. S. and BRUCE, H. M. (1962). Olfactory stimuli in mammalian reproduction. *Science, N.Y.* **134**, 1049–1054.

217. PASSANO, L. M. (1953). Neurosecretory control of moulting in crabs by the X-organ sinus gland complex. *Physiologia comp. Oecol.* **3**, 155–189.

218. PASSANO, L. M. (1960). Moulting and its control. *Physiology of Crustacea*, I. Ed. WATERMAN, T. H. Academic Press, London and New York.

219. PELLING, C. (1965). The mechanism of activation in Dipteran salivary chromosomes. II. Characteristics of puffing in giant chromosomes. *Archs Anat. microsc. Morph. exp.* **54**, 645–647.

220. PELLUET, D. and LANE, N. J. (1961). The relation between neurosecretion and cell differentiation in the ovotestis of slugs. *Can. J. Zool.*, **39**, 789–805.

221. PÉREZ-GONZÁLEZ, M. D. (1957). Evidence for hormone-containing granules in sinus glands of the fiddler crab (*Uca pugilator*). *Biol. Bull.*, Woods Hole, **113**, 426.

222. PERKINS, E. B. (1928). Colour change in crustaceans, especially in *Palaemonetes*. *J. exp. Zool.* **50**, 71–195.

223. PHILLIPS, J. E. (1964). Rectal reabsorption in the desert locust, *Schistocerca gregaria* Forskal. I. Water. *J. exp. Biol.* **41**, 15–38.

224. PIEPHO, H. (1939). Raupenhäutungen bereits verpuppter Hautstücke bei der Wachsmotte *Galleria mellonella* L. *Naturwissenschaften*, **27**, 301–302.

225. PRABHU, V. K. K. (1961). The structure of the cerebral glands and connective bodies of *Jonespeltis splendidus* Verhoeff. (Myriapoda: Diplopoda.) *Z. für Zellforsch.* **54**, 717–733.

226. RAE, C. A. (1955). Possible new elements in the endocrine complex of cockroaches. *Aust. J. Sci.* **18**, 33–34.

227. RAE, C. A. and O'FARRELL, A. F. (1959). The retrocerebral complex and ventral glands of the primitive orthopteroid *Grylloblatta campodeiformis* with a note on the homology of the muscle core of the 'prothoracic gland' in Dictyoptera. *Proc. R. ent. Soc. Lond. A.* **34**, 76–82.

228. RÖLLER, H., DAHM, K. H., SWEELY, C. C. and TROST, B. M. (1967). The structure of the Juvenile Hormone. *Angew. Chem.* **6**, 179–180.

229. ROONWAL, M. L. (1937). Studies on the embryology of the African migratory locust, *Locusta migratoria* R. and F. (Orthoptera, Acrididae). II. Organogeny. *Phil. Trans. R. Soc.* (B), **227**, 175–244.

230. SAROJINI, S. (1963). Comparison of the effects of androgenic hormone and testosterone propionate on the female Ocypod crab. *Curr. Sci.* **32**, 411–412.

231. SCHARRER, B. (1936). Über 'Drusennvenzellen' im Gehirn von *Nereis virens* Sars. *Zool. Anz.* **25**, 131–138.

232. SCHARRER, B. (1941). Neurosecretion. III. The cerebral organ of the Nemerteans. *J. comp. Neurol.* **74**, 109–130.

233. SCHARRER, B. (1941). Neurosecretion. 4. Localization of neurosecretory cells in the central nervous system of *Limulus. Biol. Bull.,Woods Hole*, **81**, 96–104.

234. SCHARRER, B. (1952). Neurosecretion. XI. The effects of nerve section on the intercerebralis-cardiacum-allatum system of the insect *Leucophaea maderae. Biol. Bull. Woods Hole*, **102**, 261–272.

235. SCHARRER, B. (1964). The ultrastructure of the corpus allatum of *Blaberus craniifer* (Blattariae). *Am. Zool.* **4**, 327–328.

236. SCHARRER, B. (1967). The neurosecretory neurone in neuroendocrine regulatory mechanisms. *Am. Zool.* **7**, 161–169.

237. SCHARRER, E. and SCHARRER, B. (1963). *Neuroendocrinology*. Columbia University Press, New York and London.

238. SCHEER, B. T. and SCHEER, M. A. R. (1951). The hormonal regulation of metabolism in crustacea. I. Blood sugar in spiny lobsters. *Physiol. comp.* **2**, 198–209.

239. SCHEFFEL, H. (1965). Der Einfluss von Dekapitation und Schnürung auf die Häutung und die Anamorphose der Larven von *Lithobius forficatus* L. (Chilopoda). *Zool. Jb. (Physiol.)*, **71**, 359–370.

240. SCHILDKNECHT, H., SIEWERDT, R. and MASCHWITZ, V. (1966). A vertebrate hormone as defensive substance of the water-beetle (*Dytiscus marginalis*). *Angew. Chem.* **5**, 421–422.

241. SCHMIALEK, P. (1961). Die Identifizierung zweier im *Tenebrio* kot und in der Hefe vorkommender Substanzen mit Juvenilhormonwirkung. *Z. Naturf.* **16b**, 461–464.

242. SCHMIALEK, P. (1963). Über die Bildung von Juvenilhormonen in Wildseidenspinnern. *Z. Naturf.* **186**, 462–465.

243. SCHNEIDERMAN, H. A. and GILBERT, L. I. (1964). Control of growth and development in insects. *Science* N.Y., **143**, 325–333.

244. SCHNEIDERMAN, H. A., KRISHNAKUMARAN, A., KULKARNI, V. G. and FRIEDMAN, L. (1965). Juvenile hormone activity of structurally unrelated compounds. *J. Insect Physiol.* **11**, 1641–1649.

245. SCULLY, U. (1964). Factors influencing the secretion of regeneration-promoting hormone in *Nereis diversicolor*. *Gen. comp. Endocr.* **4**, 91–98.

246. SIMPSON, L., BERN, H. A. and NISHIOKA, R. S. (1966). Survey of the evidence for neurosecretion in gastropod molluscs. *Am. Zool.* **6**, 123–138.

247. SLAMA, K. (1964). Hormonal control of respiratory metabolism during growth, reproduction and diapause in female adults of *Pyrrhocoris apterus* L. *J. Insect Physiol.* **10**, 283–304.

248. SLAMA, K. and WILLIAMS, C. M. (1965). Juvenile hormone activity for the bug, *Pyrrhocoris apterus*. *Proc. natn. Acad. Sci.* **54**, 411–414.

249. SLAMA, K. and WILLIAMS, C. M. (1966). The juvenile hormone. V. The sensitivity of the bug, *Pyrrhocoris apterus*, to a hormonally, active factor in American paper-pulp. *Biol. Bull.,Woods Hole*, **130**, 235–246.

250. STAMM, M. D. (1959). Estudios sobre hormonas de invertebrados. II. Aislamiento de hormonas de la metamorfosis en el ortoptero *Dociostaurus marocannus*. *Espan. Fis. Quim.* (Madrid), **55B**, 171–178.

251. STEELE, J. E. (1963). The site of action of insect hyperglycaemic hormone. *Gen. comp. Endocr.* **3**, 46–52.

252. STRANGWAYS-DIXON, J. (1961). The relationship between nutrition, hormones and reproduction in the blowfly *Calliphora erythrocephala* Meig. I. Selective feeding in relation to the reproductive cycle, the corpus allatum volume and fertilization. *J. exp. Biol.* **38**, 225–235.

253. STRANGWAYS-DIXON, J. (1961). The relationship between nutrition, hormones and reproduction in the blowfly *Calliphora erythrocephala* Meig. II. The effect of removing the ovaries, the corpus allatum and the median neurosecretory cells upon selective feeding and the demonstration of the corpus allatum cycle. *J. exp. Biol.* **38**, 637–646.

254. STRANGWAYS-DIXON, J. (1962). The relationship between nutrition, hormones and reproduction in the blowfly *Calliphora erythrocephala* Meig. III. The corpus allatum in relation to nutrition, the ovaries, innervation and the corpus cardiacum. *J. exp. Biol.* **39**, 293–306.

255. STRICH-HALBWACHS, M. C. (1954). Rôle de la glande ventrale chez *Locusta migratoria*. *C.r. Séanc. Soc. Biol.* **148**, 2087–2090.

256. STRICH-HALBWACHS, M. C. (1959). Contrôle de la mue chez *Locusta migratoria*. *Ann. Sci. nat. Zool.* Ser. 12, **1**, 483–570.

257. STUMM-ZOLLINGER, E. (1957). Histological study of regenerative processes after transection of the nervi corporis cardiaci in transplanted brains of the cecropia silkworm (*Platysamia cecropia* L). *J. exp. Zool.* **134**, 315–326.

258. TAKI, I. (1964). On the morphology and physiology of the branchial gland in Cephalopoda. *J. Fac. Fish. Anim. Husb. Hiroshima Univ.* 5, 345–417.

259. THOMAS, P. J. and BHATNAGAR-THOMAS, P. L. (1968). Use of a juvenile hormone analogue as insecticide for pests of stored grain. *Nature, Lond.* **219**, 949.

260. THOMSEN, E. (1949). Influence of the corpus allatum on the oxygen consumption of adult *Calliphora erythrocephala* Meig. *J. exp. Biol.* **26**, 137–149.

261. THOMSEN, E. (1952). Functional significance of the neurosecretory brain cells and corpus cardiacum in the female blowfly, *Calliphora erythrocephala* Meig. *J. exp. Biol.* **29**, 137–172.

262. THOMSEN, E. and HAMBURGER, K. (1955). Oxygen consumption of castrated females of the blowfly, *Calliphora erythrocephala* Meig. *J. exp. Biol.* **32**, 692–699.

263. THOMSEN, E. and MÖLLER, I. (1963). Influence of neurosecretory cells and of corpus allatum on intestinal protease activity in the adult *Calliphora erythrocephala* Meig. *J. exp. Biol.* **40**, 301–321.

264. THOMSEN, M. (1965). The neurosecretory system of the adult *Calliphora erythrocephala*. II. Histology of the neurosecretory cells of the brain and some related structures. *Z. Zellforsch.* **67**, 693–717.

265. TOYAMA, K. (1902). Contributions to the study of silkworms. I. On the embryology of the Silkworm. *Bull. Coll. Agric. Tokyo, imp. Univ.* **5**.

266. VAN DER KLOOT, W. G. (1955). The control of neurosecretion and diapause by physiological changes in the brain of the cecropia silkworm. *Biol. Bull., Woods Hole*, **109**, 276–294.

267. WEISMANN, A. (1864). Die nachembryonale Entwicklung der Musciden nach Beobachtungen an *Musca vomitoria* und *Sarcophaga carnaria*. *Z. wiss. Zool.* **14**, 187–336.

268. WELLS, M. J. and WELLS, J. (1959). Hormonal control of sexual maturity in *Octopus*. *J. exp. Biol.* **36**, 1–33.

269. WIGGLESWORTH, V. B. (1936). The function of the corpus allatum in the growth and reproduction of *Rhodnius prolixus* (Hemiptera). *Q. Jl. microsc. Sci.* **79**, 91–121.

270. WIGGLESWORTH, V. B. (1939). Häutung bei Imagines von Wanzen. *Naturwissenschaften*, **27**, 301.

271. WIGGLESWORTH, V. B. (1940). The determination of characters at metamorphosis in *Rhodnius prolixus* (Hemiptera). *J. exp. Biol.* **17**, 201–222.

272. WIGGLESWORTH, V. B. (1940). Local and general factors in the development of 'pattern' in *Rhodnius prolixus* (Hemiptera). *J. exp. Biol.* **17**, 180–200.

273. WIGGLESWORTH, V. B. (1943). The fate of haemoglobin in *Rhodnius prolixus* (Hemiptera) and other blood-sucking arthropods. *Proc. R. Soc. Series B*, **131**, 313–339.

274. WIGGLESWORTH, V. B. (1948). The functions of the corpus allatum in *Rhodnius prolixus* (Hemiptera). *J. exp. Biol.* **25**, 1–14.

275. WIGGLESWORTH, V. B. (1952). The thoracic gland in *Rhodnius prolixus* (Hemiptera) and its role in moulting. *J. exp. Biol.* **29**, 561–570.

276. WIGGLESWORTH, V. B. (1953). Determination of cell function in an insect. *J. Embryol. exp. Morph.* **1**, 269–277.

277. WIGGLESWORTH, V. B. (1954). *The physiology of insect metamorphosis*. Cambridge University Press.

278. WIGGLESWORTH, V. B. (1955). The breakdown of the thoracic gland in the adult insect, *Rhodnius prolixus*. *J. exp. Biol.* **32**, 485–491.

279. WIGGLESWORTH, V. B. (1957). The action of growth hormones in insects. *Symp. Soc. exp. Biol.* **11**, 204–227.

280. WIGGLESWORTH, V. B. (1959). *The control of growth and form*. Cornell University Press.

281. WIGGLESWORTH, V. B. (1961). Some observations on the juvenile hormone effect of farnesol in *Rhodnius prolixus* Stål. *J. Insect Physiol.* **7**, 73–78.

282. WIGGLESWORTH, V. B. (1963). The juvenile hormone effect of farnesol and some related compounds: quantitative experiments. *J. Insect Physiol.* **9**, 105–119.

283. WIGGLESWORTH, V. B. (1963). The action of moulting hormone and juvenile hormone at the cellular level in *Rhodnius prolixus*. *J. exp. Biol.* **40**, 231–245.

284. WIGGLESWORTH, V. B. (1964). The hormonal regulation of growth and reproduction in insects. *Adv. in Insect Physiology*, **2**, 247–336.

285. DE WILDE, J., DUINTJER, C. S. and MOOK, L. (1959). Physiology of diapause in the adult Colorado beetle. I. The photoperiod as a controlling factor. *J. Insect Physiol.* **3**, 75–85.

286. WILLIAMS, C. M. (1946). Physiology of insect diapause: the role of the brain in the production and termination of pupal dormancy in the giant silkworm *Platysamia cecropia*. *Biol. Bull., Woods Hole*, **90**, 234–243.

287. WILLIAMS, C. M. (1952). Physiology of insect diapause. IV. The brain and prothoracic glands as an endocrine system in the cecropia silkworm. *Biol. Bull., Woods Hole*, **103**, 120–238.

288. WILLIAMS, C. M. (1961). The juvenile hormone. II. Its role in the endocrine control of molting, pupation and adult development in the cecropia silkworm. *Biol. Bull., Woods Hole*, **121**, 572–585.

289. WILLIAMS, C. M. (1963). The juvenile hormone. III. Its accumulation and storage in the abdomens of certain male moths. *Biol. Bull., Woods Hole*, **124**, 355–367.

290. WILLIAMS, C. M. and ADKISSON, P. L. (1964). Physiology of insect diapause. XIV. An endocrine mechanism for the photoperiodic control of pupal diapause in the oak silkworm, *Antheraea pernyi*. *Biol. Bull., Woods Hole*, **127**, 511–525.

291. WILLIAMS, C. M., ADKISSON, P. C. and WALCOTT, C. (1965). Physiology of insect diapause. XV. The transmission of photoperiod signals to the brain of the oak silkworm, *Antheraea pernyi*. *Biol. Bull., Woods Hole*, **128**, 497–507.

292. YAGI, K. and BERN, H. A. (1965). Electrophysiologic analysis of the response of the caudal neurosecretory system of *Tilapia massambica* to osmotic manipulations. *Gen. comp. Endocr.* **5**, 509–526.

293. YAGI, K., BERN, H. A. and HAGADORN, I. R. (1963). Action potentials of neurosecretory neurones in the leech, *Theromyzon rude*. *Gen. comp. Endocr.* **3**, 490–495.

Index

Italicized page numbers refer to Figures